Mathematics in Rural Appalachia: Place-Based Mathematics Lessons

Edited by

Theresa M. Hopkins

&

Vena M. Long

University of Tennessee

Appalachian Collaborative Center
for Learning, Assessment, and Instruction in Mathematics
[ACCLAIM]

Koinonia Associates LLC
Clinton, Tennessee

Copyright © 2008
Theresa Hopkins.
All rights Reserved.

No part of this publication may be reproduced, stored in a retrieval system, or transmitted in any form or by any means – electronic, mechanical, photocopy, recording, or otherwise – without receiving prior permission from the publishers.

ISBN 978-1-60658-004-2

Cover photo: Dent's Run Covered Bridge - courtesy of the West Virginia Division of Tourism

Koinonia Associates LLC
Box 763
Clinton, TN 37717

Learn how you can become a published author:
www.PublishWithKA.com

Table of Contents

Introduction   1

Chapter 1: Why Rural Place-Based Lessons   3

Chapter 2: Assessing Place-Based Lessons   9

Chapter 3: The Lessons   15

    Lesson 1: The RC Cola and Moon Pie Festival   19

    Lesson 2: Traveling to Mountaineer Stadium on the PRT   25

    Lesson 3: Organizing, Summarizing, and Displaying Data from the PRT   31

    Lesson 4: Mini-Monongalia Mathematics Lesson   37

    Lesson 5: The Henry Clay Iron Furnace   43

    Lesson 6: The Mathematics of Canning   47

    Lesson 7: Wrapping and Filling   55

    Lesson 8: How Many Hay Bales?   61

    Lesson 9: How Much Hay Will Fit?   69

    Lesson 10: Estimating the Cost of a New Barn Roof   77

Lesson 11: Round Barns	85
Lesson 12: Morgantown Lock and Dam	91
Lesson 13: The Average Elevation of Decker's Creek Trail	97
Lesson 14: Are We Really Mountaineers?	103
Lesson 15: 4-H Club and Farm Animals	109
Lesson 16: Dent's Run Covered Bridge	117
Lesson 17: Rodeo Statistics for Middle School Students	123
Lesson 18: Farmer Jonah's New Barn	141
Lesson 19: How Much Will My Chicken Farm Cost?	147
Lesson 20: Confidence Intervals in Achievement Data	153
Chapter 4: Organizing, Rethinking, and Reorganizing	159

Introduction

The Appalachian Collaborative Center for Learning, Assessment, and Instruction in Mathematics (ACCLAIM), created in 2001 as a part of the National Science Foundation's Centers for Teaching and Learning, was designed to look at the issues in the intersection of mathematics education and rural education. Middle Appalachia served as the initial laboratory with all rural regions of America ultimately involved. The mission of ACCLAIM is to *"cultivate the indigenous leadership in mathematics education in rural areas."*

To meet this goal, several programs were instituted, including creation of the Appalachian Association of Mathematics Teacher Educators (AAMTE), an affiliate of the national organization AMTE, preservice teacher conferences, a doctoral program in mathematics education targeting geographically isolated individuals and extensive research in rural mathematics education.

The doctoral program, a collaboration of the Universities of Tennessee, Kentucky, Louisville, Ohio University, and West Virginia University, allows educators to remain at their schools, where they are desperately needed, while pursuing an advanced degree. With courses offered online during fall and spring semesters and residential courses in the summers, members of the doctoral cohorts receive a quality education.

Within the doctoral program is coursework in mathematics, mathematics education, rural education/sociology, and research. One manner in which this connection was manifested is in the creation of place-based lessons by cohort students and others working with ACCLAIM. This book is a collection of some of those lessons.

MATHEMATICS IN RURAL APPALACHIA

Chapter 1: Why Rural Place-Based Lessons?

Why does ACCLAIM place an emphasis on place-based pedagogy in the cohort's coursework? The primary reason is the potential positive influence on student learning when lessons have meaning to their lifeworld and the community building that occurs when schools acknowledge the culture in which they are located. However, rural teachers are at a distinct disadvantage because of the rarity of rural context in current curriculum materials.

In terms of academic achievement, several studies suggest that students involved in place-based education exhibit higher achievement test scores in comparison to other students (Loveland, 2003; Rural School and Community Trust, 2003). Loveland recounts the successes of students taking part in the Alaska Rural Systemic Initiative (AKRSI), which showed that students involved in place-based educations made greater achievement gains on the CAT-5 math tests than students not participating. In this project, students worked on a variety of issues critical to their communities, such as fishing, medicinal plants, hunting, logging, and beaver habitats.

The Rural Trust (2003), in a report reviewing six communities implementing place-based learning, found several positive outcomes. The report states "place-based learning has the power to engage and re-engage vulnerable young people in rigorous academics" (p. 5). Significant increases in test scores were reported in several of the case studies. In Virginia, schools participating in the Appalachian Rural Education Network (AREN) project found student test scores in history, computer technology, and earth sciences higher than students in comparable schools. The AREN project focused on "community media projects to address issues confronting them (the students)" (p. 3). With topics of interest to them and their community, including producing films and videos highlighting the Appalachian Underground Railroad and the Scotia Mine Disaster, it is clear to see why scores increased in history, computer technology, and earth sciences.

The claimed benefits of place-based education are not limited to increases in test scores. The Rural Trust (2003) also found increased attendance rates, lower dropout rates, increased parental and community involvement in the schools, higher aspirations and expectations for students, and, finally, an overall greater enthusiasm for learning. These benefits are certainly critical to students' achievement, but they are also important in building and maintaining community. Theobald and Nachtigal (1995) write that "Education is not just about improving standardized test scores or being first in the world' in math and science. It is also about learning to live well in a community" (p. 11).

Too often, education in rural areas is about leaving the community. Out-migration, the loss of community members with highly developed skills can have negative outcomes for the community (Smith & DeYoung, 1992). Nachtigal (1997, p. 23) agrees that the current "model of schooling...educate[s] rural students to leave their local communities, to find a job elsewhere. He continues by describing schools as a "specialized system whose purposes have been defined somewhere else, and whose interest is being served out there, somewhere" (p. 22).

How can values of place address these concerns? They can do this by creating connections between educational goals and the communities within which the schools reside. The Rural Trust (2003) suggests that the community provide the context for student learning. Connecting the community (both its people and resources) and the schools to study community needs help to build and sustain communities. Leo-Nyquist (1997, p. 3) believes "rural schools have the potential to function as both catalyst and role model for addressing a wide range of community needs and options for the future." The goal should be, as stated by Theobald and Nachtigal (1995, p. 8), "the art of living well where they are." By using the community itself to create the context for today's curriculum objectives, the importance of community becomes clear, and one's community becomes important.

Too often, rural life is ridiculed or presented in a negative light. With no direct connection between the curriculum and place, as well as the inherent stereotypes of rural place, it is no wonder that education is often viewed as a ticket to a "better" life. Place-based education can counteract this thought process by showing the students the importance of "book-learning" in their community while learning about situations unique to their place. The *Foxfire* series is a place-based project that showcases the "genius of ordinary places" (Theobald & Nachtigal, 2003, p. 11) and moves away from the idea that rural is somehow inferior. Place-based learning can create students that are "both academic achievers and good citizens: (Loveland, 2003).

To achieve this academic success and build local community, rural school curriculum must arguably be anchored in "place." Teachers in Northeastern North Carolina saw the need for place (Mahoney, 2003, p. 7), requesting "professional development opportunities based in the values and resources of the places we live-the ecology, the economy, the people, the folklore, the social structure, the history." By moving to a place-based focus, teachers can become "the producers and creators-instead of simply the *consumers* of curriculum materials and approaches that are appropriate to their own needs and contexts" (Leo-Nyquist, 1997, p. 4). With a curriculum lacking a real-world context, or with a totally non-rural context, place-based education is necessary to provide that real-world context according to many observers (e.g. Yager, 2003).

This compilation of place-based lessons will not remove the challenge of creating a real-life community-based context for students of mathematics. With the variety of communities within the Appalachian region alone, such a compilation would require many volumes. The goal of a book like this is to provide some examples to help teachers begin to look to their communities and their mathematics pedagogy differently. Perhaps some of the lessons will translate directly. Others may work with a slight modification with respect to their state curriculum objectives or community. Our main aim, however, is to spark an idea for place-based curriculum and pedagogy anchored to the culture in which your students live.

Organization of the Book

This book is a compilation of lessons contributed primarily by students in the second ACCLAIM doctoral cohort, with an additional lesson or two contributed by authors who have been connected with the program at various times. Most lessons are at the middle or high school level, but with minor adjustments can be taught at other grade levels. Keeping in mind the need to make adjustments responsive to your place, space is left in the text for readers to jot notes and make comments.

We created a specific layout with which to present each lesson. The introduction provides some background information about the lesson and its "place."

Following this introduction there is a 4-section box highlighting the NCTM Content and Process Standards addressed in that particular lesson (NCTM, 2000). The third box lists the state standards (from the state of place). The final block is left for you to add standards specific to your place.

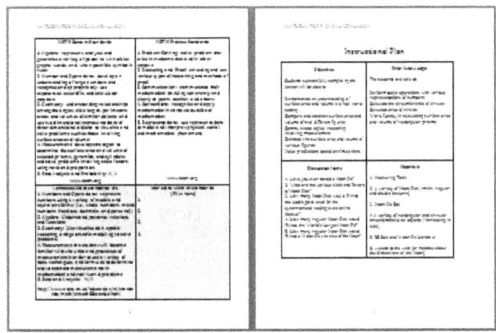

The lessons continue with the Instructional Plan, which includes the lesson objectives and the prior knowledge necessary for successful completion of the lesson.

Lesson Procedures follow, with Discussion Items and Questions on which to focus during the lesson as well as the materials needed for implementation. These are followed by the lesson itself, often with suggestions for extension and assessment.

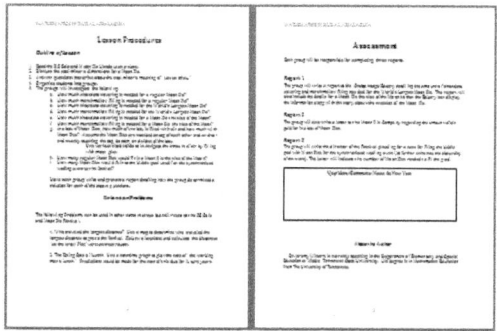

After the Instructional Plan and Lesson Procedures, an empty box is included for you to write your ideas, comments, or notes for next year. Does the context need adjusting to better fit the circumstances found in Bulls Gap Tennessee, as opposed to Morgantown, West Virginia? Does your school size or lay-out require an adjustment to how the lesson is implemented? Perhaps the lesson was interesting and generated some ideas for an extension or entirely new lesson. This is the place to record your ideas.

Finally, we include a brief author bio for each lesson. The lesson creators are "real-folk" working in real schools, just like you are doing, facing many of the same issues you are-presenting state-required curriculum, being held accountable for students' test scores, and struggling to make the material connect with the students' lives. Perhaps you might want to email one of the authors to find out more about the lesson or just chat about similar issues you might be facing.

References

Leo-Nyquist, D. and Theobald, P. (1997). *Towards a pedagogy of place: Finding common ground for rural researcher teacher educators, and practitioners.* Roundtable notes prepared for the1997 American Educational Research Association, Chicago. (ERIC Document Reproduction Service No. ED432416)

Loveland, E. (2003). Achieving academic goals through placed-based learning: Students in five states show how to do it. *Rural Roots,* 4(1), 1 and 6-8.

Mahoney, C. R. (2003). *Mathematics education in rural communities: A mathematician's view.* Presented at the 2002 Appalachian Collaborative Center for Learning, Assessment, and Instruction in Mathematics Research Symposium, Macarthur, OH.

Nachtigal, P. (1997). Place value: Experiences from the rural challenge. In *Coming home: Developing a sense of place in our communities and school.* Proceedings of the 1997 Forum. (ERIC Document Reproduction Service No. ED421311)

National Council of Teachers of Mathematics. (2000), *Principles and standards for school mathematics.* Reston, VA: Author.

Rural School and Community Trust. (2003). *Engaged institutions: Impacting the lives of vulnerable youth through place-based learning.* (ERIC Document Reproduction Service No. ED481278)

Smith, E. D. and DeYoung, A. J. (1992). *Exploratory studies of occupational structure of the workforce and support of public education in rural Appalachia.* Mississippi State: Southern Rural Development Center. (ERIC Document Reproduction Service No. ED392586)

Theobald, P. and Nachtigal, P. (1995). Culture, community, and the promise of rural education. *Phi Delta Kappan 77*(2), 132-135.

Yager, R. (2003). Place-based education: What rural schools need to stimulate real learning. *Rural Roots, 4*(1), 9.

Chapter 2: Assessing Place-Based Mathematics Lessons

Place-based lessons can be used to introduce, extend, or practice many mathematical skills. These lessons will result in a variety of cognitive and non-cognitive outcomes, some of which are intended and some which are not. This section outlines a process to help teachers (1) think about and identify important outcomes from place-based mathematics lessons, (2) develop strategies to assess these outcomes informally and/or formally, (3) use the results of the assessment to guide future instruction, and (4) document and share the results with students and others.

Determining Lesson Outcomes

Good teachers constantly assess students informally. Both formal and informal assessments can be embedded in instruction providing important information for the instructional decisions made by teachers. Formal assessments are stronger if they relate closely to the content and form of the classroom instruction. Place-based mathematics lessons provide excellent opportunities for assessment but also create potential pitfalls. Assessments must be clearly aligned with the important mathematics targeted by the lesson and must not veer into assessing the context rather than the mathematics.

The lessons in this book are clearly designed to ensure that students learn and apply important mathematics. Some important outcome questions with regard to mathematics include:

- What mathematics skills and concepts do students need to complete the activities in the lesson?
- What mathematics skills and concepts might students learn as a result of completing the activities in the lesson?
- What higher-order mathematics processes (problem solving, reasoning, inferring, applying) are involved in the activities of the lesson?
- How might students communicate their knowledge of mathematics during the lesson?
- The contextual nature of placed-based mathematics lessons often requires connecting mathematics learning to other disciplines like science, social studies, language arts, music, or art. As a result, cognitive outcomes related to these disciplines may emerge during lessons. While this knowledge is important and valuable to the student, assessing this knowledge may lead to diluting the mathematical focus.

Likewise, the cultural nature of some place-based lessons means that students will likely need or learn knowledge about their place and/or culture. In designing lessons some important questions about outcomes with regard to place and culture include:

- What knowledge of place or culture do students need in order to complete the activities of the lesson?
- What might students learn about place or culture as a result of completing the activities of the lesson?
- What strategies might students use to learn about their place or culture in the activities of the lesson?

The place-based mathematics lessons in this book lend themselves to group work and other cooperative activities. As a result, non-cognitive factors like teamwork, cooperation, motivation, or attitude are often important outcomes in the activities. Some important questions about non-cognitive student outcomes might include:

- How important is teamwork and cooperation in completing the activities of the lesson?
- What important beliefs and attitudes, about mathematics and about place, should surface as a result of completing the activities of the lesson?
- How important is motivation and persistence in completing the activities of the lesson?

Specific answers to these questions will help teachers ensure powerful learning. Furthermore, the clearer teachers are about their expectations of student performance and outcomes the better chance that the assessments developed provide useful information.

Sometimes, particularly with new lessons, teachers may be sure about the intent of the lesson but unsure about the likelihood of achieving those outcomes. In this case, smaller, formative assessments can provide teachers information about student outcomes during the activities. Teachers can then use the results of these on-going assessments to refine the lesson to more precisely achieve the desired learning.

Prioritizing Outcomes

Place-based lessons like the ones in this book will generate a range of varied outcomes. Once teachers generate their lists of outcomes, the next step is to prioritize the outcomes. During this process, they may ask the following questions:

- What is the critical mathematics content to be learned as a result of the activities of the lesson?
- What other content is to be learned through the activities of the lesson?
- What should students learn about place or culture through the activities of the lesson?

- What should students learn about themselves through the activities of the lesson?

As teachers grapple with these questions, they should focus on what is really important to them and their students. Prioritizing the important outcomes can lead to assessments that provide valuable information to teachers. Trying to assess too many outcomes or trying to assess too many outcomes that are not mathematical can weaken the assessment process and provide information about students that is vague or irrelevant.

Developing an Assessment Strategy

The next step in creating a viable assessment plan for place-based lessons is to develop a strategy for assessing important outcomes. If these lesson(s) constitute only a small fraction of the overall chapter being taught, informal embedded assessment may be most appropriate. If the nature of the unit of instruction is largely contextual, then the summative assessment of the unit should mirror the nature of the instruction. Assessment strategies can range from closed tasks like true-false, multiple-choice, or fill-in-the-blank to more open tasks like open- or constructed-response questions to larger open tasks like presentations, papers, posters, artwork, or portfolios. The individualist, highly contextual nature of these lessons makes open assessments more appropriate. For example, the activities in most of these lessons require challenging applications of mathematics content, thoughtful exploration of cultural phenomena, and analysis of complex data. Therefore, performance events would serve as the optimal assessment strategy. In developing an assessment strategy, teachers are encouraged to use a variety of assessment tasks. Smaller tasks may be used to assess progress during the lessons, while larger performance assessments may be used as culminating assessments. Teachers also may use traditional tests and quizzes to assess mastery of important skills and then performance events to assess application and reasoning.

Developing Strategies to Evaluate Student Outcomes

Once teachers have determined their assessment strategies, they must then decide how to evaluate the quality of the student performance on the assessments. Traditional assessments like tests and quizzes can be scored and graded in traditional ways. More open types of assessments like constructed-response and performance events will require more complex evaluation strategies. The most common strategy for evaluating student performance on these types of assessments is a rubric, which assigns values to different levels of student performance. Rubrics can either be general or specific. That is, a general rubric can be developed to assess a class of tasks, like all presentations or project. Specific rubrics are designed to assess performance on a specific assessment task. A general four point rubric suitable for low stakes, classroom assessment might be as follows:

    4 – To die for...
    3 – Totally acceptable
    2 – Needs help
    1 – Knock Knock??

Given the idiosyncratic nature of place-based activities, specific rubrics may be more appropriate. In that case each level of assessment would be described in terms of the task used for the assessment.

Sample Assessment Task from *RC Cola and Moon Pie Festival*

Working in pairs, **each** of you designs a product to compete with Moon Pies. Individually describe your product in terms of surface area and volume. Working together, compare and contrast the surface area and volume of your two products. Write a rationale to convince a company to produce one of your products.

To be turned in:
- *1 paper each, prepared individually, describing your product.*
- *1 paper, prepared together, comparing and contrasting your products and the rationale to convince a company to produce one of your products.*

Rubric:
    4   All work is complete, all calculations are accurate, work shows individual and joint efforts, rationale uses convincing and accurate logic.
    3   All work is complete and any errors are casual. Work shows some evidence of individual and joint effort. Rationale is acceptable but could be clearer and more compelling. With prompting the work could be improved without further instruction on the part of the teacher.
    2   Work is incomplete. Errors show the need for further instruction.
    1   Problem is attempted but no progress in the right direction is evident.

Once a rubric has been selected or developed, teachers then have the choice of using the rubric themselves to evaluate student work or asking students or others to use the rubric to evaluate student work. The latter strategy has the advantage of helping all students focus on the attributes of performance and more clearly understanding expectations for such performance.

Assessment Resources

Teachers who desire more assistance in developing and implementing a viable assessment system for their place-based mathematics lessons can refer to the following publications:

Angelo, T. A., & Cross, K. P. (1993). *Classroom assessment techniques A handbook for college teachers.* San Francisco: Jossey-Bass

Bush, W. S. (ed.) (2000). *Mathematics assessment: Cases and discussion questions for grades 6-12.* Reston, VA: National Council of Teachers of Mathematics.

Bush, W. S., & Leinwand, S. (eds.) (2000). *Mathematics assessment: A practical handbook for grades 6-8.* Reston, VA: National Council of Teachers of Mathematics.

Bush, W. S., & Greer, A. S. (eds.) (1999). *Mathematics assessment: A practical handbook for grades 9-12.* Reston, VA: National Council of Teachers of Mathematics.

Depka, E. (2001). *Designing rubrics for mathematics.* Glenview, IL: Pearson Professional Development.

Johnson, D. W., & Johnson, R. T. (2004). *Assessing students in groups: Promoting group responsibility and individual accountability.* Thousand Oaks, CA: Corwin Press.

National Council of Teachers of Mathematics. (1995). *Assessment Standards for School Mathematics.* Reston, VA: The Council.

National Council of Teachers of Mathematics (2005). *Mathematics Assessment Sampler: Grades 6-8.* Reston, VA: The Council.

Popham, W. J. (1998). *Classroom assessment: What teachers need to know (2nd edition).* Boston: Allyn and Bacon.

Stenmark, J. K. (ed.) (1991). *Mathematics assessment: Myths, models, good questions, and practical suggestions.* Reston, VA: National Council of Teachers of Mathematics.

Stiggins, R.. K. (1994). *Student-centered classroom assessment.* New York: Merrill.

Wilcox, S. K., & Lanier, P. E. (2000). *Using assessment to reshape mathematics teaching: A casebook for teachers and teacher educators, curriculum and staff development specialists.* Mahwah, NJ: Lawrence Erlbaum Associates.

MATHEMATICS IN RURAL APPALACHIA

Chapter 3: The Lessons

There are twenty place-based lessons in this initial volume. Six states are represented: Ohio, Kentucky, Missouri, North Carolina, Tennessee, and West Virginia. The fact that the majority of these lessons are centered on West Virginia should not be surprising as cohort members were attending classes at the University of West Virginia when asked to create a place-based lesson. While the initial focus of the lesson might be at the middle or high school method, or designed for future mathematics teachers, with some changes, all lessons can be altered and taught at a variety of levels.

The book starts with a lesson we can all sink our teeth into: *The RC Cola and Moon Pie Festival*. This lesson looks at the humble beginnings of the Moon Pie and the yearly festival surrounding Moon Pie and RC Cola in the town of Bell Buckle, Tennessee. Jeremy Winters's description of his family's visit makes the festival come to life and creates a great context for several state mathematics standards (representing numbers in a variety of equivalent forms, understanding patterns, using geometric modeling to solve problems, and, becoming familiar with the units and processes of measurement).

Following Jeremy's lesson is our first lesson based in West Virginia, Morgantown, to be exact. The Personal Rapid Transit (PRT) running between downtown Morgantown and various spots on the campus of the University of West Virginia is the context for Jamie Fuggitt's lesson involving measurement and volume. The lesson focuses on the acceptable size of bags/purses/backs which riders can take to one of the PRT's destinations: The University of West Virginia's football stadium.

Sherry Jones's data-based lesson follows. Sherry's lesson, like Jamie's, uses the Morgantown PRT for context. However, rather than a focus on the volume of packages, Sherry's lesson has students collecting data involving the number of people using the PRT over the previous 12 months and using statistical software to organize and display the data.

Jeremy Zelkowski's *Mini-Monongalia Math Lesson* is the first of two lessons centered on the Henry Clay Iron Furnace. Jeremy's lesson uses Law of Sines to estimate distances in addition to other measurement activities involving the weight of a stone in the building of the furnace and the volume of various solids. Paula Schlesinger's lesson involving the Henry Clay Iron Furnace takes a more open ended approach. Paula provides her students with various facts and figures related to the furnace and asks for the students to develop their own questions and investigations.

Paula adds another place-based lesson next, *The Mathematics of Canning*. Although Paula's lesson hails from North Carolina, canning is a practice used widely in many areas. Her lesson focuses on different mathematical topics, including looking at the linear relationship between the volume and mass of green beans canned. This 3 day series of plans involves students gathering canning information the first day, estimating and predicting the amount of cans from 2 peck baskets of green beans on day two, concluding with actual canning on the third day.

Victor Brown, a resident of Kentucky, follows with *Wrapping and Filling* which focuses on measurement of surface area and volume using different units of measurement. By providing contexts of farming issues, such as area for planting corn-seed or setting tobacco, repainting the barn or other outbuildings, building a temporary silo, Victor anchors the mathematics in place.

Carolyn Best and Sherry Jones follow with two lessons related to hay. Carolyn, in *How Many Hay Bales?* Looks at the different loading styles for round bales so that they can be transported where they are needed. Students investigate different ways to load hay bales to maximize the amount per delivery. Sherry's lesson, *How Much Hay Will Fit?* has students calculating the capacities of different structures for holding hay bales. Students are not only calculating the volumes of irregularly shaped buildings, but developing estimating and problem-solving skills (if the hay bales have a certain dimension, how does that change the amount of hay that can be stored).

The collection continues with another barn-related lesson from Johnny Belcher. With agriculture playing a large part in the history a West Virginia, Johnny developed a lesson surrounding the round barn design which was touted for its large loft space. Geometry skills involving calculating area, surface area, and volume, and representing three dimensional figures as well as measurement skills are all a part of Johnny's lesson. Wouldn't it be great to follow Sherry's lesson with Johnny's? Is the round barn truly advantages in terms of hay storage?

In keeping with the barn theme, Courtenay Mayes has developed a lesson which has students investigating the cost of building a barn roof for a successful horse owner in Kentucky. Her lesson includes a variety of mathematics such as the study of slope and pitch, 3-dimensional visualization, calculation of surface area, use of the Pythagorean Theorem, and finally, calculation of material costs. This investigation is a multi-day event.

In keeping with the study of slope, *Morgantown Lock and Dam*, submitted by Ron Smith, has students investigating the history of locks and dams on the Monongahela River and collecting data from previous years (such as river depth, amount of cargo delivered, and the depth of the water in the lock while the level changes over time) to investigate the idea of slope.

Michael Ratliff follows with a lesson that has students collect the elevation at various points along Decker Creek Trail in West Virginia (or supplies the data points if an actual field trip is not a possibility). The lesson will include how to collect data points to answer the question "what is the average elevation of Decker's Creek?" the number of data points to collect, using a sample of data points, etc.

Are We Really Mountaineers? has students comparing the elevation of the different ranges in the Appalachian Mountains. Nicolyn Smith's lesson has students collecting data such as the names/locations of the ranges, tallest mountain in each range (name and height in meters) then calculating a ratio and percentage of the height of each of these mountains to the highest peak in the world.

Nicolyn continues with another lesson near and dear to many students' hearts in Ohio and elsewhere: participation in the 4-H club and the raising and care of farm animals. Within the context of 4-H, students calculate percent, fraction, and decimal equivalents using data from the Feeder Calf Project Animal Inventory.

Dents Run Covered Bridge is based on one of only 17 covered bridges left in West Virginia. Students will be involved with creating scale drawings and models of bridges in this multi-day lesson. Bridges are tested for length, height, and strength. The author, Debbie Waggoner, suggests having a civil engineer come to your classroom to discuss bridges existing in your area.

Traveling rodeos provide the context for the next two lessons from Jamie Fugitt. Jamie has students studying measures of center, outliers and spread (including range and interquartile range) with data collected from the Barrel Racing and Bull Riding at the 2003 and 2004 Forsyth Rodeo. Students will look at different representations for the data as well as make conclusions about the data from representations.

Farmer Jonah's New Dairy Barn is designed to facilitate middle school mathematics teachers' use of and skill with the *Geometer's Sketchpad* (Key Curriculum Press) while emphasizing important mathematics. Michael Ratliff's lesson involves investigating, with the Sketchpad, the best location for a new dairy barn on Farmer Jonah's farm, located north of County Road #421.

In keeping with the farming slant, Sharilyn Owens offers a mathematics/business lesson involving the cost of operating a chicken farm. Students will create a scale drawing of their own land (or grandparent's or neighbor's) as well as construction costs and limitations as well as earnings for different chicken coop sizes.

Finally, Paula Schlesinger offers a lesson involving statistics that should be of interest to teachers and students everywhere. Her lesson, entitles simply *Confidence Intervals* investigates why some states included confidence intervals in conjunction with the No Child Left Behind Act of 2001 (NCLB). The statistical analysis will be connected to rural school issues within the home state.

We hope that some of these descriptions have sparked your interest enough to give them a try, or perhaps has created some ideas of your own that are specific to your place. Enjoy!

Lesson 1

The RC Cola & Moon Pie Festival

By Jeremy Winters

A Brief History

According to southern legend, the Moon Pie received its name in the following way. In the early 1900's, a Chattanooga baker was looking for a new item to bake in his shop. He decided to go and ask some of the local coal miners what they liked to snack upon. One coal miner told the baker that his favorite treat was to take two graham crackers and put marshmallows in-between. As the moon was rising, the baker asked the coal miner how big this treat should be. The coal miner put his hands into a circle outlining the moon and proclaimed, "As big as the moon and just as thick!" The baker, inspired by the idea, created the Moon Pie.

In the 1950's, the popularity of the Moon Pie was at an all time high. The cost of the Moon Pie at this time was five cents. Blue-collar workers of the day would use the inexpensive snack as lunch. Needing a drink to accompany the Moon Pie, the workers reached for the biggest and cheapest drink available, RC Cola. Thus, the duo of the RC Cola and Moon Pie became known as "the working man's lunch." The total cost of the lunch was 15 cents. A Moon Pie and an RC Cola continue to be one of the South's most popular pairs.

Celebrating the Famous Duo

Each June since 1995, the town of Bell Buckle, Tennessee pays homage to this southern favorite. Travelers from all over the United States come to witness the town's festivities.

The festival includes a parade where local honorees throw Moon Pies to those lining the streets. Vendors at the festival sell Moon Pies and RC Colas in a variety of forms. A craft fair accompanies the festival where you can purchase lovely items made by local artisans. One such craft at the fair this year was a purse made out of an original Moon Pie packaging box. . The festival is capped off with the cutting of the world's largest Moon Pie.

Contests and Prizes

MATHEMATICS IN RURAL APPALACHIA

The festivities include a ten-mile run, clogging, a parade, bluegrass music, synchronized wading in a kiddy pool (This is a riot!), games, and prizes. At the main stage, contests go on throughout the day. Some of the contests included at the festival are facts and trivia about the RC Cola and Moon Pie, facts and trivia about the town of Bell Buckle, youngest and oldest person at the festival, and who traveled the farthest to get to the festival.

The culminating event of the festival is the cutting of the World's Largest Moon Pie. To protect the famous pie from melting in the heat, the Moon Pie is kept in cool conditions at the Chamber of Commerce. This year, the largest Moon Pie had a diameter of 34 inches and a depth of 3 inches. As I was walking around, I overheard the stage DJ's give the dimensions of the largest Moon Pie ever as 4 ½ feet in diameter and 6 inches thick.

The World's Largest Moon Pie

NCTM Content Standards	NCTM Process Standards
Algebra: represent, analyze, and generalize a variety of patterns with tables, graphs, words, and, when possible, symbolic rules; Number and Operations: develop an understanding of large numbers and recognize and appropriately use exponential, scientific, and calculator notation; Geometry: understanding relationships among the angles, side length, perimeters, areas, and volumes of similar objects; and use two dimensional representations of three-dimensional objects to visualize and solve problems such as those involving surface area and volume; Measurement: develop strategies to determine the surface area and volume of selected prisms, pyramids, and cylinders; and solve problems involving scale factors, using ratio and proportion; Data Analysis and Probability: N/A www.nctm.org	Problem-Solving: solve problem that arise in mathematical and in other context; Reasoning and Proof: selecting and use various types of reasoning and methods of proof; Communication: communicate their mathematical thinking coherently and clearly to peers, teacher, and others; Connections: recognize and apply mathematics in contexts outside of mathematics; Representations: use representations to model and interpret physical, social, and mathematical phenomena www.nctm.org
Tennessee State Standards 6-8 Numbers and Operations: represent numbers using a variety of models and equivalent forms (i.e., whole numbers, mixed numbers, fractions, decimals, and percents); Algebra: Understand patterns, relations, and functions; Geometry: Use visualization, spatial reasoning, and geometric modeling to solve problems; Measurement: the student will become familiar with the units and processes of measurement in order to use a variety of tools, techniques, and formulas to determine and to estimate measurements in mathematical and real-world problems Data and Analysis: N/A http://www.state.tn.us/education/ci/standards/math/cimath68stand.shtml	**Your State/District Standards (fill in here)**

Instructional Plan

Objectives	Prior Knowledge
Students successfully completing the lesson will be able to: • Demonstrate an understanding of surface area and volume in a real world setting; • Compare and contrast surface area and volume of two different figures; • Communicate logical reasoning involving measurement; • Estimate the surface area and volume of various figures; • Make predictions based on simulations.	The students are able to: • Perform basic operations with various representations of numbers; • Calculate the circumference of circles; • Calculate area of circles; • Work fluently in calculating surface area and volume of rectangular prisms.
Discussion Items	**Materials**
• Have you ever tasted a Moon Pie? • What are the various sizes and flavors of Moon Pies? • How many Moon Pies would fit into the kiddy pool used for the synchronized wading event at the festival? • How many regular Moon Pies would fit into the World's Largest Moon Pie? • How many regular Moon Pies would fit into a Moon Pie the size of the Moon?	• Measuring Tools • A variety of Moon Pies (minis, regular, and double deckers) • Moon Pie Box • A variety of rectangular and circular three-dimensional objects (increasing in size) • RC Cola and Moon Pie Handout • Access to the web (or handout about the dimensions of the Moon)

Lesson Procedures

1. Read the RC Cola and Moon Pie Handout as a class.
2. Discuss the coal miner's dimensions for a Moon Pie.
3. Answer questions that arise about the coal miner's meaning of "just as thick."
4. Organize students into groups.
5. The groups will investigate the following.
    a. How much chocolate covering is needed for a regular Moon Pie?
    b. How much marshmallow filling is needed for a regular Moon Pie?
    c. How much chocolate covering is needed for the World's Largest Moon Pie?
    d. How much marshmallow filling is needed for the World's Largest Moon Pie?
    e. How much chocolate covering is needed for a Moon Pie the size of the Moon?
    f. How much marshmallow filling is needed for a Moon Pie the size of the Moon?
    g. In a box of Moon Pies, how much of the box is filled with air and how much with Moon Pies? Assume the Moon Pies are stacked on top of each other and end-to-end exactly touching the top, bottom, and sides of the box.
    h. Use various sized solids to investigate the amount of air by filling with moon pies.
    i. How many regular Moon Pies would fit in a Moon Pie the size of the Moon?
    j. How many Moon Pies would fit into the kiddy pool used for the synchronized wading event at the festival?

Have each group write and present a report detailing how the group determined a solution for each of the above questions.

Extension Problems

The following Problems can be used in other content areas but still relate to the RC Cola and Moon Pie Festival.

1. Who traveled the longest distance? Use a map to determine who traveled the longest distance to get to the festival. Pick two locations and calculate the distances "as the crow flies" versus street routes.

2. The Rising Cost of Lunch. Use a cost-time graph to plot the cost of "the working man's lunch." Predictions could be made for the cost of this duo for future years.

Assessment

Each group will be responsible for completing three reports.

Report 1

The group will write a report to the Chattanooga Bakery detailing the amount of chocolate covering and marshmallow filling needed for the World's Largest Moon Pie. The report will also include the details for a Moon Pie the size of the Moon so that the Bakery can display the information along with the story about the creation of the Moon Pie.

Report 2

The group will also write a letter to the Moon Pie Company regarding the amount of air paid for in a box of Moon Pies.

Report 3

The group will write the director of the Festival pleading for a case for filling the kiddy pool with Moon Pies for the synchronized wading event (to further enhance the absurdity of the event). The letter will indicate the number of Moon Pies needed to fill the pool.

Your Ideas/Comments/Notes for Next Year

About the Author

Dr. Jeremy Winters is currently teaching in the Department of Elementary and Special Education at Middle Tennessee State University. His degree is in Mathematics Education from The University of Tennessee.

Lesson 2

Traveling to Mountaineer Stadium on the PRT

By: Jamie Fugitt

Introduction

The Personal Rapid Transit (PRT) is the primary mode of transportation between Downtown Morgantown and various locations on West Virginia University campuses. During the semester, on an average school day, 71 cars shuttle an average of 16,000 people between the various locations. During Mountaineer Week each year, student organizations participate in a PRT cram where they attempt to see how many people can cram into a PRT car. The record is somewhere around 100.

The PRT system is used extensively to shuttle people between locations on days when the Mountaineers have a home football game. For safety and security reasons certain sized items are not allowed in the stadium. The sign is posted at the Beechhurst PRT station so that when boarding the PRT people will be aware of the items which should not be brought to the stadium. According to the sign, bags, purses, packs, etc, which are brought into the stadium can "not exceed 12' x 12' x 12'."

PRT photos by WVU Photographic Services

NCTM Content Standards	NCTM Process Standards
Number and Operations: Compute fluently and make reasonable estimates. Algebra: Use mathematical models to represent and understand quantitative relationships. Geometry: Use visualization, spatial reasoning, and geometric modeling to solve problems. Measurement: Understand measurable attributes of objects and the units, systems, and processes of measurement; and Apply appropriate techniques, tools, and formulas to determine measurements. Data Analysis and Probability: N/A www.nctm.org	Problem-Solving: Solve problems that arise in mathematics and other contexts. Reasoning & Proof: Make and investigate mathematical conjectures. Communication: Communicate their mathematical thinking coherently and clearly to peers, teachers, and others Connections: Recognize and apply mathematics in contexts outside of mathematics Representations: Use representations to model and interpret physical, social, and mathematical phenomena www.nctm.org
West Virginia State Standards (6th grade) • Solve problems, in context, involving operations on whole numbers, fractions, and decimals. • Investigate and model volume and surface area. • Apply formulas to determine perimeter, circumference and/or area of plane figures. http://wvde.state.wv.us/csos/	**Your State/District Standards (fill in here)**

Instructional Plan

Objectives	Prior Knowledge
Students successfully completing the lesson will be able to: - Explain the need for precision of mathematical statements; - Critically analyze the meanings of a statement involving measurements; - Determine the volume of a rectangular prism; - Estimate the volume of various three dimensional objects; - Physically measure some dimensions and calculate other measurements on various three - dimensional objects; and - Provide logical reasoning for making a decision involving measurements.	The students are able to: - Define and calculate surface area - Measure with standard and metric units. - Define and calculate volume - Perform basic operations with whole numbers, fractions, and decimals - Find the area of plane figures - Represent 3-D figures - What is the PRT?
Discussion Items	**Materials**
- What does it mean for a bag not to "exceed 12" x 12" x 12"?" - Is volume or surface area a better representation? Why? - What do you think is the best mathematical model for meeting the stadium requirements?	- Measuring tools - Bags of various sizes - Precise wording of the regulations on the size bags that can be used. - Poster - Pizza box - Umbrella - Soda Bottle

Lesson Procedures

May I Take My Bag to the Stadium?

Post a picture of the sign and allow students to discuss in groups various interpretations of what it means for a bag to "exceed 12" x 12" x 12"."

Allow each group to share one interpretation of the meaning. Possible interpretations might include:

> The volume of the bag does not exceed 1728 in^3.
> The girth of the bag does not exceed 48 in.
> No side of the bag has a surface area which exceeds 144 in^2.
> The total surface area of the bag does not exceed 864 in^2.
> The sum of the three dimensions of the bag does not exceed 36 in.
> The length plus the girth does not exceed 60 in.
> No dimension on the bag exceeds 12 in.

Others – encourage students to provide other interpretations.

Provide some bags and ask students to determine if they will be allowed in the stadium under the various interpretations of the rule.

Provide each group with an object, such as a rolled up poster, a large pizza box, a soccer ball, a 2 liter soda bottle, an umbrella, etc, and ask students in the groups to determine if the object can be placed in a carrying container and then taken into the stadium. Have the students determine which interpretations of the sign (if any) will allow the item to be brought into the stadium. Have students sketch the design of a bag which would be allowed in the stadium and which would hold their item.

Many more interesting and worthwhile mathematical tasks related to the PRT could be posed. Some of these could be posed and answered from the information provided on the front page of this activity. For other mathematical tasks it might actually be necessary to collect data related to the cars' arrival times at various stations, the number of people boarding at certain stations, the time of day people rides the PRT, etc.

Assessment

Ask each student to write a letter to the Mountaineer Stadium manager explaining the ambiguity of the information on the sign. Each student should also suggest which interpretation they believe is best and provide rationale for their decision.

A FUN CHALLENGE: Determine the volume of the inside of a PRT car. This would require researching information about the cars. It might be necessary for the teacher to gain permission for the students to be allowed to measure the inside dimensions of a car. Answer the question, "Suppose 100 students crammed into a PRT car. How much space would be available for each student?"

Your Ideas/Comments/Notes for Next Year

About the Author

In addition to completing a doctoral degree through ACCLAIM and being employed full-time at College of the Ozarks, where she serves as Chair of the Division of Mathematical and Natural Sciences, Jamie Fugitt enjoys spending time with her family. Her husband Jeff is also a doctoral student pursing a degree in religious studies and the social sciences. Their son Johnathan is 21 and graduated in May from William Jewell College in Kansas City. Elizabeth, their daughter, is 19 and is a sophomore at Missouri State University.

Jamie can be reached at fugitt@cofo.edu

Lesson 3

Organizing, Summarizing, and Displaying: Data from the PRT in Morgantown, WV

By: Sherry Jones

Introduction

The Personal Rapid Transit (PRT) described in the prior lesson, *Traveling to the Mountaineer Stadium on the PRT*, offers a variety of mathematical investigation possibilities. While the previous lesson had a geometry and measurement focus, this lesson provides students an opportunity to explore data analysis issues.

According to the NCTM *Standards*, collecting, representing, and processing data are necessary skills in society. This lesson will consolidate, deepen and build on exploratory data analysis developed in previous grades. If the teacher chooses, students may use software to construct the graphs and perform statistical calculations. However, it is critical that students have an understanding of which measures are appropriate for answering certain question and how to interpret the graphical representations.

PRT photos by WVU Photographic Services

NCTM Content Standards	NCTM Process Standards
Number and Operations: Compute fluently and make reasonable estimates. Algebra: Understand patterns, relations, and functions; analyze change in various contexts Geometry: n/a Measurement: n/a Data Analysis and Probability: Formulate questions that can be addressed with data and collect, organize, and display relevant data to answer them; select and use appropriate statistical methods to analyze data; develop and evaluate inferences and predictions that are based on data. www.nctm.org	Problem-Solving: Solve problems that arise in mathematics and other contexts. Reasoning & Proof: Make and investigate mathematical conjectures. Communication: Communicate their mathematical thinking coherently and clearly to peers, teachers, and others Connections: Recognize and apply mathematics in contexts outside of mathematics Representations: Use representations to model and interpret physical, social, and mathematical phenomena www.nctm.org
West Virginia State Standards (Algebra) Collect, organize, interpret data and predict outcomes using the mean, mode, median, and range. Analyze a given set of data for the existence of a pattern numerically, algebraically and graphically; determine the domain and range; and determine if the relation is a function. http://wvde.state.wv.us/csos/	**Your State Standards (fill in here)**

Instructional Plan

Objectives	Prior Knowledge
Students will be able to: • Create graphical representations of information collected over a period of time • Find the mean, median, and mode, variance and standard deviation of a data set • Recognize trends that appear in the data • Discover types of questions that can be answered by the data set and its graphical representation • Make inferences based on the statistics and graphs generated	The students have: • Background knowledge of the PRT • Basic operational skills
Discussion Items *See Instructional Plan*	**Materials** • Data gathered from the Personal Rapid Transit (PRT) Control Room in Morgantown, WV • Graph Paper or software such as Excel • Calculators • Pencil and Paper

Lesson Procedures

According to the NCTM *Standards*, collecting, representing, and processing data are necessary skills in society. This lesson will consolidate, deepen and build on exploratory data analysis developed in previous grades.

If the teacher chooses, students may use software to construct the graphs and perform statistical calculations. However, it is critical that students have an understanding of which measures are appropriate for answering certain question and how to interpret the graphical representations.

Divide students in small groups and distribute data that has been gathered over the previous 12 months regarding number of travelers on the PRT in Morgantown, WV.

- Ask students to organize the data in some meaningful way and represent the data graphically.
    - For example, the students could organize the data by month and produce a histogram or line chart showing the average number of travelers per month.

- Ask students to perform statistical calculations either by hand or using a calculator or software.
    - They should use their graphical representation to further interpret the data to spot trends and shape of the data set.

- Discourse should be encouraged about which type of graphical representations will help answer certain questions.
    - For example, a line graph may be more useful to some students in interpreting trends where a histogram may be more useful in interpreting shape of the data and months of peak usage of the PRT.

- Students should be encouraged to see the relationships between different types of graphical representations.
    - How are they alike and how are they different?

- Students should interpret what the data shows and what it does not show. They should be encouraged to pose questions and determine the answers from the data and its representations.

Possible Questions for Students to Explore

1. What do the mean, median, and mode tell us about the data set? Are the three values the same? If not, why are they different? Which measurements are affected by extremes or outliers in the data set?

2. What might explain outliers in the data set? What might explain peak usage times?

3. How does the data vary from the mean?

4. How do the various statistical measurements relate to the graphical representations of the data?

5. What might account for any trends that are noted in the data set?

6. What types of decisions about how the PRT operates might be based on the data and its graphical representations?

7. What is the average number of travelers per day over the last 12 month period?

Assessment

Ask each group to share their graphical representations and finding with the class. A rubric for grading each presentation could be developed and/or students could perform self and peer assessments. Any assessment should consider the level of understanding and connections demonstrated by the group.

Your Ideas/Comments/Notes for Next Year

About the Author

MATHEMATICS IN RURAL APPALACHIA

Sherry Jones completed her doctorate in 2008 and is an Associate Professor of Business Education at Glenville State College in Glenville, WV. She has taught math-oriented courses in the business department at Glenville for 18 years. In 2005, Sherry was awarded the Curtis Elam Professor of Teaching Excellence Award.

Prior to teaching at Glenville State, Sherry taught upper level mathematics courses at Gilmer County High School in Glenville, WV, for seven years. She was recognized as a Gilmer County Teacher of the Year for her efforts in this position.

Sherry has lived in Gilmer County all of her life. She and her husband, David, raised two sons, Chris and Casey Jones. Chris and his wife, Amy, have given Sherry what she describes as great joy in her life: two grandsons, Orrin and Elijah Jones.

Sherry can be reached at Sherry.Jones@glenville.edu

Lesson 4

Mini-Monongalia Mathematics Lesson

By: Jeremy Zelkowski

Introduction

This lesson encompasses Cooper's Rock State Forest. It is the largest state park in West Virginia, including a breath taking scenic overlook of the Cheat River. The Cheat River feeds the Monongahela River, which forms the Ohio River along with the Alleghany River in Pittsburgh, PA. Cheat Lake, a man-made lake, feeds the Cheat River. What's amazing is that the Cheat River and Monongahela River merge to become the Mon River, is the only truly dedicated northern flowing river in the northern hemisphere of the world. There are very small rivers that share some of these properties; however, there are none that have the volume and flow rate of the Mon River.

In Cooper's Rock State Forest, there is also the Henry Clay Iron Furnace. The Henry Clay Iron Furnace is a history lesson in itself. Hence, we have mathematics, physical education, and even history embedded within the lesson.

NCTM Content Standards	NCTM Process Standards
Number and Operations: Compute fluently and make reasonable estimates. Algebra: Use mathematical models to represent and understand quantitative relationships. Geometry: Use visualization, spatial reasoning, and geometric modeling to solve problems. Measurement: Understand measurable attributes of objects and the units, systems, and processes of measurement; and Apply appropriate techniques, tools, and formulas to determine measurements. Data Analysis and Probability: N/A www.nctm.org	Problem-Solving: Solve problems that arise in mathematics and other contexts. Reasoning & Proof: Make and investigate mathematical conjectures. Communication: Communicate their mathematical thinking coherently and clearly to peers, teachers, and others Connections: Recognize and apply mathematics in contexts outside of mathematics such as physical education and history education. Representations: N/A www.nctm.org
West Virginia State Standards (9-12 grade)	**Your State/District Standards (fill in here)**
Solve problems, in context, involving operations on whole numbers, fractions, and decimals. Investigate and model volume and weight. Apply formulas to determine volume of three dimensional figures. Formulate questions that can be addressed with data and collect, organize, and display relevant data to answer them. http://wvde.state.wv.us/csos/	

MATHEMATICS IN RURAL APPALACHIA

Instructional Plan

Objectives	Prior Knowledge
Students successfully completing the lesson will be able to: - Estimate the distance by using the Law of Sines; - Critically estimate the weight of one stone used in the building of the furnace; - Estimate the volume of various three dimensional objects; - Physically measure some dimensions and calculate other measurements on various three dimensional objects; and - Provide logical reasoning for making a decision involving measurements.	The students are able to: - Measure with standard and metric units. - What is the Iron Furnace? - Define and calculate volume - How pig iron is produced? - Find the weight of 3-D shapes
Discussion Items	**Materials**
- What the volume of the Henry Clay Iron Furnace? - What is the weight of one stone used in the building of the furnace? - The furnace is said to have produced 4 tons of pig iron per day. This is a rate of change. Suppose the furnace produced 1.5% more pig iron each month or year. About how much total pig iron was produced for the 11 years the furnace was in use?	- A 50' tape measure - Twelve inch protractor - Two foot square (right angle) - Fish scale (up to 15 lbs) - Twenty feet of twine - Two dollars in quarters

Lesson Procedures

On a Saturday, or even a normal school day, a field trip to Cooper's Rock State Forest would be to gather the data for the lesson. Students would form groups, and be given the following tools for gathering the data: A 50' tape measure, a 12 inch protractor, a 2 foot square (right angle), a fish scale (up to 15 lbs), 20 feet of twine, and $2.00 in quarters. There assignment for the day would be the following:

> Estimate the distance from the Scenic overlook to the rock structure which is an obvious site on the other side of the cheat river.

> Find the volume of the Henry Clay Iron Furnace. Also, estimate the weight of one stone used in the building of the furnace.

> Record any pertinent data from the historical marker at the furnace site. You will research the history of iron furnaces in the area during the mid-1800's and write a small paper and give a presentation to the class.

> The P.E. lesson would simply be the exercise, since the assignment for the day would end up with about 3-4 miles of hiking. There could also be a small race to the furnace and/or on the return. This trek is about 1 to 1.5 miles each way, downhill on the way, and uphill the way back. Students could also record their times for each way. This could later be a math lesson on statistics, and also be used to try to find the declination and inclination of the trek.

Assessment

Ask each student to measure the angle of ascent on each side, the building block height, find a piece of the actual stone and record it's weight and dimensions to estimate the weight of a big block, and more.

A Fun Challenge

Record any pertinent data from the historical marker at the furnace site. You will research the history of iron furnaces in the area during the mid-1800's and write a small paper and give a presentation to the class.

Your Ideas/Comments/Notes for Next Year

About the Author

Jeremy Zelkowski received his BA and MS in Mathematics from West Virginia University. Born and raised in Wheeling, WV, Jeremy has been a mathematics instructor at Louisiana School for Math, Science, and Arts and is currently a Senior Lecturer of Mathematics at WVU.

Jeremy is a member of the second ACCLAIM cohort and completed his doctorate through Ohio University. His research interest include mathematics instruction at the undergraduate level, undergraduate mathematic curriculums and course design (Calculus and Pre-calculus courses) and the confounding factors which MAY predict mathematics success and degree completion for college students.

Lesson 5

The Henry Clay Iron Furnace

By: Paula Schlesinger

The Henry Clay Iron Furnace is located in the Cooper's Rock State Forest on the Cheat River northeast of Morgantown, WV. Henry Clay Furnace, located on Quarry Run was built between 1834 and 1836 by Leonard Lamb for Tassey and Bissell. It was a cold blast furnace and produced 4 tons of pig iron each 24 hours. It was one of several furnaces that were operated in this area during the nineteenth century and was used until about 1847. About 200 people were employed at the furnace; it was the center of a community of over a hundred dwellings with a store, church and schoolhouse. Ownership of the furnace was conveyed in 1839 to the Ellicott's who built a system of wooden-railed tramways that ran through the mountains connecting the furnaces and ore pits until 1845. All of the iron produced was floated down the Cheat River. The pattern of industrial development is constantly changing. The iron industry cycle on Cheat Mountain is now complete.

This lesson is for 6th to 8th grade. The lesson is very open-ended, asking the students to make up their own word problems.

MATHEMATICS IN RURAL APPALACHIA

NCTM Content Standards	NCTM Process Standards
Content standards will vary due to the open-ended nature of this activity. Specific standards will depend upon the direction the students take with their problem creation www.nctm.org	Process standards will vary due to the open-ended nature of this activity. Specific standards will depend upon the direction the students take with their problem creation www.nctm.org
West Virginia State Standards **6th grade** State standards will vary according to problems generated by students. http://wvde.state.wv.us/csos/	**Your State Standards** **(fill in here)**

Objectives	Prior Knowledge
Mathematical objectives for this lesson will vary based on its very open-ended nature. Students will investigate data, conjecture solutions, and communicate their ideas to classmates	While prior mathematical knowledge will be dependent upon the path the students follow when creating their specific problem. Certainly, prior experience with open-ended problems would be necessary for students to have an acceptable comfort level when participating in the lesson.
Discussion Questions See next page for key questions to close the lesson.	**Materials** - 8 ½" X 11" construction paper - Scotch tape - Masking tape - Popped popcorn - Markers - Rulers - Centicubes and/or multilink cubes - inch^3 cubes

Instructional Plan

Lesson Procedures

Share with the class the picture of the Henry Clay Iron Furnace and the information given at the beginning of the lesson. The lesson is very open-ended, asking the students to make up their own word problems. Some possible topics would be:

The environmental impact of the iron furnace: How many acres of forest were cut down to produce the charcoal to run the furnace during the 12 years it was in production?

How many cords of wood were used in a year to make charcoal for the furnace in a year?

Are the facts given about the number of acres of trees used in a year and the amount of charcoal needed to produce a ton of iron and the pig iron production of the furnace consistent?

The economic impact of the furnace: If we assume an average price of $35 per ton and transportation costs of $10 per ton, how much revenue was produced by the furnace per year?

Estimate the height, width and volume of the furnace.

Estimate the number of blocks that were used to build the furnace.

About the Author

Paula Schlesinger is a doctoral student in the ACCLAIM program. She teaches mathematics and German at Mayland Community College in Spruce Pine, NC. Paula grew up on a farm in northern Indiana, and has four children (who are mostly grown) and one grandchild.

MATHEMATICS IN RURAL APPALACHIA

Lesson 6

The Mathematics of Canning

By: Paula Schlesinger

Introduction

The lesson centers on a common rural activity—canning green beans. Many families in rural communities keep vegetable gardens in spring and summer and "put up" the produce for use during the winter months. Children who have helped their mothers and grandmothers can probably do not realize how much mathematics is involved in this activity.

As the MSEB says "Mathematics is a natural mode of human thought…" (p. 44). Canning involves measuring, estimating and predicting, and ratio and proportion. This lesson involves the physical activity of weighing the green beans before and after preparation. It connects mathematics and science, because low acid vegetables (such as green beans) need to be processed at 240ºF in order to prevent contamination by botulism. The students will need to learn about the dangers of botulism and what effect the elevation of their community and the application of pressure have on the boiling point of water.

NCTM Content Standards	**NCTM Process Standards**
Number and Operations: develop, analyze, and explain methods for solving problems involving proportions. Algebra: use symbolic algebra to represent situations and to solve problems, especially those that involve linear relationships Geometry: use geometric models to represent and explain numerical and algebraic relationships Measurement: use common benchmarks to select appropriate methods for estimating Data Analysis and Probability: make conjectures about possible relationships between two characteristics on the basis of scatter plots www.nctm.org	Problem Solving: solve problems that arise in mathematics and other contexts. Reasoning and Proof: make and investigate mathematical conjectures. Communication: organize and consolidate their mathematical thinking through communication. Connections: recognize and apply mathematics in contexts outside of mathematics Representation: use representations to model and interpret physical phenomena www.nctm.org
North Carolina **Grade 7** Solve problems involving volume and surface area of cylinders, prisms, and composite shapes. Identify, analyze, and create linear relations, sequences, and functions using symbols, graphs, tables, diagrams, and written descriptions. Develop flexibility in solving problems by selecting strategies and using mental computation, estimation, calculators or computers, and paper and pencil. http://www.ncpublicschools.org/curriculum/mathematics/scos/	**Your State Standards** **(fill in here)**

Instructional Plan

Objectives	Prior Knowledge
Students successfully completing the lesson will be able to: - Discover the linear relationship between volume and mass, - Convert measures of volume. - Make connections to science (relationship between boiling point of water and pressure)	- Estimation and Prediction - Volume - Data Collection - 3-Figures (Cylinders) - Units and Unit Conversions - Best Fit Line
Key Questions	**Materials**
- What supplies are needed to can green beans? - What temperature is recommended for processing canned beans? - Why do the beans need to be processed at this temperature? - What is the elevation of our community? - What is the boiling point of water at our elevation? - How do you achieve the temperature required for processing beans?	- Approximately 2 pecks of green beans - Scales, - Knives - Cutting boards - Canning jars - Water pitcher - Colander - Zip lock bags - Kitchen

Lesson Procedures

The Mathematics of Canning - Day 1: Research

Divide the students into groups and give each group a research question. Then allow the students to go on the internet or look up the answers to their questions in the school learning resources center. The questions they should be given are:
- What supplies are needed to can green beans?
- What temperature is recommended for processing canned beans?
- Why do the beans need to be processed at this temperature?
- What is the elevation of our community?
- What is the boiling point of water at our elevation?
- How do you achieve the temperature required for processing beans?

The students should be able to use a search engine, such as Google to find the required information, but may need some help from the teacher about what key words to use. Here are some websites that have the information:
- A step-by-step guide for canning just about anything sponsored by Ball Corporation, a manufacturer of canning jars: www.homecanning.com
- Several state universities have links to the United States Department of Agriculture home canning guide. One such site is: http://foodsafety.cas.psu.edu/canningguide.html
- A site that has a boiling point calculator is http://www.biggreenegg.com/boilingPoint.htm

After the students have completed their research, there should be a class discussion on the dangers of botulism and safety issues involved in canning. For example, the importance of washing hands, sterilizing jars and lids and processing beans at the proper temperature and for the proper length of time.

The home economist for the county extension service could also be invited to talk to the class about safety issues.

Students should volunteer to bring in equipment from home. You may have to purchase new bands and lids. If you are borrowing a pressure canner, be sure to have it checked out by the county home economist. You may be able to borrow a canner from the county extension service, if you cannot borrow one from one of the students' families.

The Mathematics of Canning - Day 2: Estimation and Prediction

Equipment and supplies needed: approximately 2 baskets of green beans (approximately 1 peck each), measuring scales, knives, cutting boards, canning jars, water pitcher, colander, zip lock bags.
>Again, divide the students into groups of 3 or 4.
>>Each group will get from 0.5 to 1.5 pounds of beans.
>>They should weigh the beans and record the weight

They should also make an estimate of how many ounces they think their beans will make once they have been prepared and record that amount.

When the entire contents of one basket have been used, make a note of that; so that there is a record of the number of pounds of beans in a peck. If the entire content is not used, weigh the beans remaining after all the groups have gotten their beans.

Now each group should get a cutting board and one or two knives and prepare the beans: snap off the ends, string, and cut into one-inch pieces.

After the beans are prepared, pack them into a canning jar and cover with water so that the water just covers the beans. If the jar is full, be sure to leave one-inch head room.

Record the number of ounces of prepared beans (use canning jars that have a graduated scale in ounces). Dump the contents of the jar into a colander, then pack the beans into zip-lock bags to be refrigerated until the next class, when the beans will be canned.

Now, the class as a whole will work together to make a scatter plot with the number of pounds of beans each group started with on the horizontal axis and the number of quarts of prepared beans on the vertical axis (for most groups this will be less than one quart). Start by listing the data for each group on the chalk board. Then the class should discuss what scale should be used on each axis. After the axes have been marked off, each group should place their dot on the plot.

- Does this relationship appear to be linear? Estimate the line of best fit for the data, and sketch it in on the scatter plot.
- Use this line to estimate how many pounds are needed to make one quart of prepared beans.
- Answer the following questions:
- How close was each group's estimate to the actual amount of beans produced? Which group had the closest estimate?
- How many pounds of beans has the class already prepared?
- How many more pounds will we have to prepare tomorrow to make 7 quarts of beans – a standard amount for a pressure canner load?
- How many pounds are in one peck of beans?
- How many quarts can be produced from one bushel of beans?

- *Joy of Cooking* (Rombauer 1997) estimates 15 – 20 quarts per bushel of beans. How does our estimate compare to theirs?

The Mathematics of Canning - Day 3: Canning

If the school kitchen is not available for this project, you may need to make a field trip to a local church or community center for the actual canning. Parents, grandparents or other adults from the community will be helpful for the actual canning.

Make sure that you have gathered all the necessary equipment: tongs, ladle, canning funnel, jars, lids and bands, a pressure canner, cutting board, knives, scale, and a large pot to cook the beans.

Carefully follow the step-by-step instructions provided by the USDA or www.homecanning.com .

Having experienced home canners as a resource will make the project go much more smoothly.

Your Ideas/Comments/Notes for Next Year

About the Author

Paula Schlesinger is a doctoral student in the ACCLAIM program. She teaches mathematics and German at Mayland Community College in Spruce Pine, NC. Paula grew up on a farm in northern Indiana, and has four children and one grandchild.

The Mathematics of Canning - Student Handout

1. How close was each group's estimate to the actual amount of beans produced? Which group had the closest estimate?

2. How many pounds of beans has the class already prepared?

3. How many more pounds will we have to prepare tomorrow to make 7 quarts of beans, a standard amount for a pressure canner load?

4. How many pounds are in one peck of beans?

5. How many quarts can be produced from one bushel of beans?

6. *Joy of Cooking* (Rombauer 1997) estimates 15 – 20 quarts per bushel of beans. How does our estimate compare to theirs?

Lesson 7

Wrapping and Filling

By: Victor Brown

Introduction

Both surface area and, in particular, volume are central mathematical ideas that must be understood in order to make informed decisions when living in a rural setting. Whether one is computing the area of a field for planting field-corn seed, setting tobacco, or ordering lime; undertaking the project of re-painting the barn or some other out-building; ordering new 2 inch by 10 inch by 12 foot long oak planks from the local saw mill for the new surface for the bridge over the creek; building a temporary silo for this year's bumper corn crop; or simply building a fence that will keep the deer out of the kitchen garden, one must have a thorough understanding of surface area and volume so they can be a savvy rural consumer.

NCTM Content Standards	NCTM Process Standards
Number and Operations: Compute fluently and make reasonable estimates. Algebra: N/A Geometry: Use visualization, spatial reasoning, and geometric modeling to solve problems. Measurement: Understand measurable attributes of objects and the units, systems, and processes of measurement; and Apply appropriate techniques, tools, and formulas to determine measurements. Data Analysis and Probability: N/A www.nctm.org	Problem-Solving: Solve problems that arise in mathematics and other contexts. Reasoning & Proof: Make and investigate mathematical conjectures. Communication: Communicate their mathematical thinking coherently and clearly to peers, teachers, and others Connections: Recognize and apply mathematics in contexts outside of mathematics Representations: Use representations to model and interpret physical, social, and mathematical phenomena www.nctm.org
Kentucky State Standards **8th grade** Students will estimate, compare, and convert units of measures for length, weight/mass, and volume/capacity within the U.S. customary system and within the metric system: 1. length (e.g., parts of an inch, inches, feet, yards, miles, millimeter, centimeter, kilometer); 2. weight/mass (e.g., pounds, tons, grams, kilograms); and 3. volume/capacity (e.g., cups, pints, quarts, gallons, milliliters, liters). (The intent of this standard is for students to make ballpark comparisons and not to memorize conversion factors between U.S. and metric units.) http://www.education.ky.gov	**Your State/District Standards** **(fill in here)**

Instructional Plan

Objectives	Prior Knowledge
<ul><li>Understand that surface area is a measure of *wrapping* an object and volume is a measure of *filling* an object;</li><li>Estimate the volume of a variety of cylinders;</li><li>Invent personal strategies and formulas for finding the surface area and volume of objects;</li><li>Investigate the effects, given certain specifications, on surface area and volume when varying the dimensions of rectangular prisms and cylinders (This type of problem solving investigation is strongly supported by the Technology Principle in the *Principles and Standards for School Mathematics* (*PSSM*);</li><li>Look for and discover patterns in the volumes of cylinders, cones, and spheres;</li><li>Recognize and solve problems involving surface area and volume</li></ul>	The students are able to:<ul><li>Explore the concept of volume with centicubes and $inch^3$ cubes;</li><li>Using a given number of cubes to create rectangular prisms with a variety of surface areas;</li><li>Discovering the least and the greatest surface area.</li></ul>
Discussion Questions	**Materials**
See next page for key questions to close the lesson.	1. 8 ½" X 11" construction paper 2. Scotch tape 3. Masking tape 4. Popped popcorn 5. Markers 6. Rulers 7. Centicubes 8. Multilink cubes 9. $inch^3$ cubes

Lesson Procedures

Remind students of the "Check 3B - 4B". (This means that the group member with a question/concern is responsible for "checking" with at least three (3) other resources, before "checking" with Mr. Brown...see attached classroom/group management routine.)

Have the students come to consensus as a group and record their conjectures. Once a conjecture is made, one way to validate it is to create both models. If constructed accurately and neatly, the taller one would have a radius of 1.35" and a height of 11" and the shorter, squattier model would have a radius of 1.75" and a height of 8 ½".

One could choose to place the taller cylinder "inside" the shorter one and fill the taller up with the popped popcorn (or one of the other materials on hand). Then, to easily compare their volumes, simply pinch and lift the taller one to leave all the popcorn in the shorter model. If the popcorn is even with the top of the shorter, then the two volumes are about equal. If the popcorn spills over the top of the shorter, then the taller had the greater volume. If the popcorn does not even fill-up the shorter cylinder, then the shorter has the greater volume. (The reality is that even though the two cylinders have the same surface area (93.5 inches squared), the shorter cylinder (81.6 inches cubed) has almost a 30% increase in volume over the taller cylinder (62.7 inches cubed).)

The real world connection here is that one of the reasons silos are built a bit on the "squatty side" is to yield a greater volume with the same amount of material used to construct the silo sides. (Another reason is that as the height of the silo increases, the sidewall pressure on the silo greatly increases. Exploring this factor makes for a nice extension! A second extension is to explore the ratio of silo height to silo radius, because it is important that the silo not be too "squatty", for if it is the corn will rot and spoil because of a lack of aeration.) Students individually summarize their experiences and, on a separate half sheet of paper, write clarification questions to be asked/discussed in the closing meeting.

Closing Meeting Outline/Discussion Starter/Guiding Questions/BIG Ideas:
- How did the various groups' initial conjectures compare with their final conclusions?
- What was the range of answers for the volume of the taller silo?
- What was the range of answers for the volume of the shorter, squattier silo?
- What were some of the strategies that groups tried that did not work?
- What were at least three different strategies that did work?
- Have misconceptions in students thinking to determine volume been clarified?
- Who still needs help/assistance with this?
- What formula(s), whether student created or prior knowledge, were used?
- How do they "fit" with our "Copy, Plug, Chug, Label" strategy?

Cylinders with the same side surface area can have a **great difference** in their volume!!! Consider sharing/discussing several examples from the attached reference table.

Assessment

Evidence of Learning/Homework/Open Response/Attached Assessment

You are permitted to use scissors to cut the 8 ½" by 11" paper into "strips", holding the paper in either landscape or portrait position, and only using either whole numbers or "very friendly fractions" for the height of your newly created cylinders. Please construct at least two additional cylinders from separate sheets of construction paper. Find the side surface area, the area of the base, and the volume of each of your new cylinders.

Open Response Question: Farmer Brown plans to build a silo 5.2 m high to hold 80 meters cubed of corn.
- If the silo were cylindrical, what would the diameter of the base be? (4.43 meters)
- If the silo were a square based, rectangular prism, how wide would it be? (3.92 meters)
- What area of metal would be needed to make the side of the silos in Part A?...in Part B? (72.4 meters squared & 81.54 meters squared, respectively)

Your Ideas/Comments/Notes for Next Year

About the Author

Victor B. Brown has over 20 years experience teaching middle school/high school mathematics in urban, suburban, and rural school districts in Indiana, Kentucky, and Ohio. In addition, he has over 5 years experience representing a major K-8 mathematics publishing company as their National Mathematics Consultant, has served 2 years as a Kentucky Highly Skilled Educator in support of systemic school improvement in Eastern Kentucky, and has 5 years experience at the district level as a mathematics consultant/coach, in support of their implementation of a K-12 *Standards-Based/Reform Mathematics Curricula*. Currently, Victor is enrolled in the ACCLAIM doctoral program in the Department of Teaching and Learning at the University of Louisville, Kentucky, while serving as a Regional Coordinator of the Kentucky Center for Mathematics, managing the Demonstration/Training Site at Eastern Kentucky University. Victor's professional activities include Past President of the Kentucky Council of Teachers of Mathematics, professional development provider at 4 International, 6 National, and over 20 Regional conferences, planner and presenter for mathematics in-service activities in over 40 states, and in these countries: Bermuda, Germany, Columbia, Paraguay, Brazil, Argentina, England, and Italy. He is a member of NCTM, KCTM, AMTE, and AAMTE.

Lesson 8

How Many Hay Bales?

By: Carolyn Best

Introduction

Hay production in Tennessee in 2003 was 4.88 million tons according to the Tennessee Agricultural Statistics Service (http://www.nass.usda.gov/tn/ff082003.pdf). It is difficult to drive for very far in the state without seeing fields with baled hay, farms with barns full of hay, trucks delivering hay, or heavy machinery baling hay. With approximately 2.5 tons produced per acre, even small farms have a lot of hay to bale and transport.

This lesson looks at two different bale types and issues involved with safely transporting the maximum amount of hay by studying packing issues, volume of the bale-types, regulations involving the transportation of the hay in a variety of scenarios.

NCTM Content Standards	NCTM Process Standards
Number and Operations: Compute fluently and make reasonable estimates. Algebra: N/A Geometry: Use visualization, spatial reasoning, and geometric modeling to solve problems. Measurement: Understand measurable attributes of objects and the units, systems, and processes of measurement; and Apply appropriate techniques, tools, and formulas to determine measurements. Data Analysis and Probability: N/A www.nctm.org	Problem-Solving: Solve problems that arise in mathematics and other contexts. Reasoning & Proof: Make and investigate mathematical conjectures. Communication: Communicate their mathematical thinking coherently and clearly to peers, teachers, and others Connections: Recognize and apply mathematics in contexts outside of mathematics Representations: Use representations to model and interpret physical, social, and mathematical phenomena www.nctm.org
Tennessee Framework (6-8) Solve problems, compute fluently, and make reasonable estimates Analyze characteristics and properties of two- and three-dimensional geometric figures Understand measurable attributes of objects and the units, systems, and processes of measurement Apply appropriate techniques, tools, and formulas to determine measurements http://www.state.tn.us/education/ci/standards/math/7math.shtml	**Your State Standards** **(fill in here)**

Instructional Plan

Objectives	Prior Knowledge
Students successfully completing the lesson will be able to: • Explain the need for precision of mathematical statements; • Critically analyze the meanings of a statement involving measurements; • Determine the volume of a rectangular prism; • Estimate the volume of a cylinder; • Provide logical reasoning for making a decision involving measurements	The students are able to: • Define and calculate surface area • Define and calculate volume • Find the area of plane figure • Represent 3-D figures • What is the PRT? • Measure with standard and metric units • Perform basic operations with whole numbers, fractions, and decimals.
Discussion Items	**Materials**
How many of you have passed a field with baled hay? How was it baled? What shapes? How many of you have seen hay being transported?	Information about regulations available at http://www1.agric.gov.ab.ca/$department/deptdocs.nsf/all/for4856 Transparency of Loading Styles for Round Bales Graph paper

Lesson Procedures

Introducing the Activity Each pair of students will need several right circular cylinders to represent bales of hay and a copy of Student Sheet 1: Loading Styles for Round Bales. To make the right circular cylinders, provide each pair of students with graph paper, scissors, tape, and Study Sheet 1.

Display a transparency of the <u>Loading Styles for Round Bales</u> and the pattern and picture of a box.

Give them the following instructions:

- We're going to do some problems to help us find a way to predict the number of bales that will fit in a box for each loading style. Here is a picture of the box. First, predict how many bales fit in the pictured box using the first loading style. Write your prediction in the blank.
- Next, draw the box pattern on graph paper [*hold up a sheet,*] **cut** it out, fold it, and tape the edges to make the box, leaving the top of the box open. Fill your box with bales to check your prediction, and then write down the actual number of bales that fit in the box. Be sure you check your prediction for each loading style before going on the next problem. That way, you'll get better at predicting as you go along.
- Go ahead and start on the box now. Predict, build the box, and check the count with bales. Then go on to the remaining loading styles.
- Let's suppose that you are loading a truck with round bales of hay that need to be hauled to a buyer 310 miles away.
- Suppose the size of the round hay bale is 5 ft diameter by 5 ft width.
- The diagrams of the loading styles are shown on Student Sheet 2. Start with the first: crosswise. Let's load a tractor semi-trailer with round hay bales. The dimensions of the trailer that will be loaded with bales are 50 ft long by 10 ft wide.

Show the diagram on page 7.

- You want to haul as many bales of hay as possible, so you pack the hay as tightly as possible. There can only be 2 layers of bales, therefore what could be the height of the box?
- Your first job loading the truck is to solve this problem:
    How many bales of hay can be loaded using the crosswise loading style onto the semi-trailer as shown?

Assessment

Once you have found a way to predict the number of bales, you will need to explain your method in writing. That way, other people loading hay bales can use your method too.

If the height of the "box" were a consideration, would another style be more efficient?

```
┌─────────────────────────────────────────────────────────────┐
│              Your Ideas/Comments/Notes for Next Year         │
│                                                              │
│                                                              │
│                                                              │
│                                                              │
│                                                              │
│                                                              │
│                                                              │
│                                                              │
└─────────────────────────────────────────────────────────────┘
```

About the Author

Caroline Munn Best, who was the first ACCLAIM student to complete her dissertation, was born in Atlanta, GA on June 2, 1950 and was raised in Cartersville, GA where she attended public school. She graduated from Maryville College, Maryville, TN with a B.A. in Mathematics with certification to teach grades 7-12. Her senior thesis at Maryville was on teaching 8th mathematics to the slow learner. After graduation, she taught middle school mathematics for one year in Carterville, GA and 2 ½ years at Carter Middle School, Knox County, TN.

Upon moving to Chattanooga midyear, she began coursework on a master's degree in education with a concentration in mathematics at the University of Tennessee, Chattanooga. During this time, she taught developmental mathematics part-time at Chattanooga State Technical Community College. Upon graduation in 1977, she taught mathematics as an adjunct at UTC and also supervised the mathematics lab for developmental mathematics students. After moving to Maryville, TN and raising three children, she returned to full-time teaching in 1990 at Pellissippi State Technical Community College where she is currently coordinator of developmental mathematics.

MATHEMATICS IN RURAL APPALACHIA

Student Sheet 1

How Many Bales?

How many bales fit in the box? Predict. Then build a box and use your bales to check. Check your prediction before going on to the next style. Think about a way you could predict the number of bales that would fit in any box.

Style	Picture	Prediction	Actual
Crosswise			
Pipe Style			
Mushroom Style 1			
Mushroom Style 2			
Barrel Style			

1. Which style allows more bales to be loaded on the trailer?
2. Are there other advantages that one style has over another that would influence which one is chosen?

Student Sheet 2

Loading Styles for Round Bales

Crosswise	
Pipe style	
Mushroom Style 1 (lower layer stands on end)	
Mushroom Style 2 (lower layer stands on end)	
Barrel Style (both layers stand on end)	

Further Investigation

1. The permit that you must obtain requires the following:
   - No more than one third of any bale can extend beyond the edge of the trailer deck or hayrack.
   - For any truck or trailer, the number of bales loaded on an upper layer shall not exceed the number of bales loaded on the lower layer.

   How do these rules change the most efficient style?

   Would another style become more efficient?

2. What if the size of the bale was changed, for instance, the diameter remains 5 ft but the length is 6 ft?

   Does this change the most efficient style?

   Would another style become more efficient?

3. Are rectangular bales for efficient for hauling? Find out the dimensions of a rectangular bale and compare the volume of a round and rectangular bale. Then determine the most efficient style of loading rectangular bales onto the same size trailer.
   How does the number of rectangular bales compare to the number of round bales?

4. If rectangular bales are more efficient, then why are round bales even used? Talk to a farmer and found out the advantages of round bales of hay versus rectangular.

Lesson 9

How Much Hay Will Fit?

By: Sherry Jones

Introduction

Farmer Jones is still using the same barn that his great grandfather used to store hay. A picture of the old barn is included above. Farmer Jones wants to store square bales of hay in the loft of the barn but cannot remember exactly how many bales of hay will fit in the loft. He needs to know how many bales the loft will hold because he may have to find extra storage for his hay in another location. The hay crop this year promises to be a good one.

A square bale of hay is not really square but rectangular in shape. Farmer Jones' baler produces bales of hay that measure 14 inches by 18 inches by 36 inches. The bales of hay are bound by two long pieces of twine, each of which wraps around the entire hay bale as noted in the following picture.

In order to help Farmer Jones, you will need to know the dimensions of the loft.

The picture and drawing below will help you with this. The floor of the loft measures 12 feet by 18 feet.

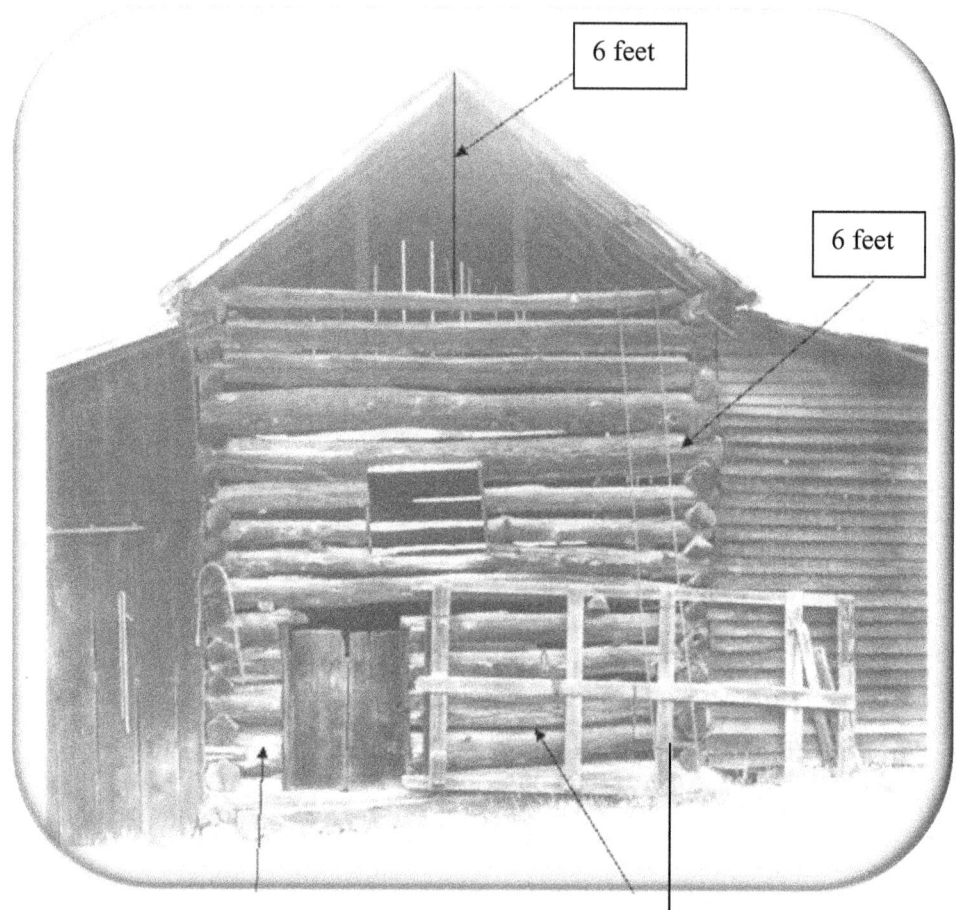

The lower level of the barn is used to shelter cows and horses and cannot be used to store hay. Only the upper middle part of the barn is used for hay storage because the horses and cows will be fed in stalls on the sides of the barn. The hayloft is shaped as in the following drawing. This view is as if you were looking down on the hayloft from above.

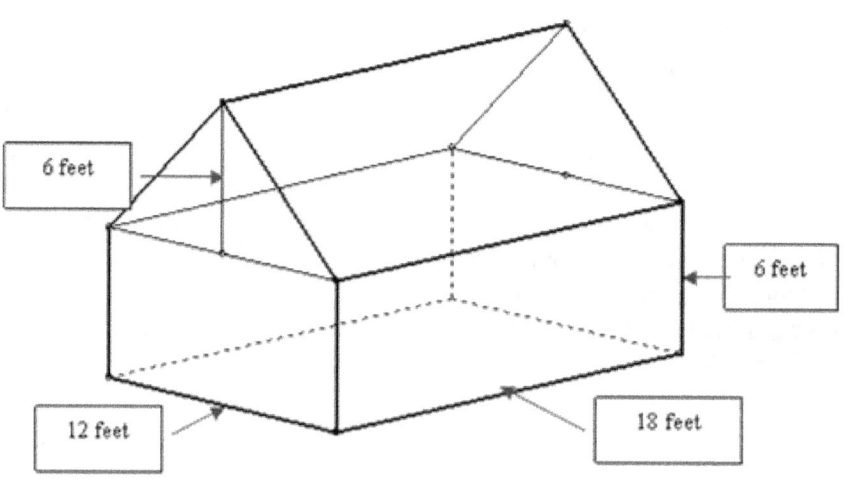

NCTM Content Standards	NCTM Process Standards
Number and Operations: Compute fluently and make reasonable estimates. Algebra: Use mathematical models to represent and understand quantitative relationships. Geometry: Use visualization, spatial reasoning, and geometric modeling to solve problems. Measurement: Understand measurable attributes of objects and the units, systems, and processes of measurement; and Apply appropriate techniques, tools, and formulas to determine measurements. Data Analysis and Probability: N/A www.nctm.org	Problem-Solving: Solve problems that arise in mathematics and other contexts. Reasoning & Proof: Make and investigate mathematical conjectures. Communication: Communicate their mathematical thinking coherently and clearly to peers, teachers, and others Connections: Recognize and apply mathematics in contexts outside of mathematics Representations: Use representations to model and interpret physical, social, and mathematical phenomena. www.nctm.org
West Virginia State Standards **7th grade** Solve application problems with whole numbers, decimals, fractions and percents Use appropriate estimation strategies in problem situations including evaluating the reasonableness of a solution. Use 2-dimensional representations of 3-dimensional objects to visualize and solve problems Use the concept of volume for prisms, pyramids, and cylinders as the relationship between the area of the base and height http://wvde.state.wv.us/csos/	**Your State Standards** **(fill in here)**

Instructional Plan

Objectives	Prior Knowledge
Students successfully completing the lesson will be able to: • Utilize appropriate operations and tools to estimate and calculate the number of rectangular solids that will fit in a given space; • Utilize logical reasoning and problem solving strategies to answer application questions	The students are able to: • Define and calculate surface area • Define and calculate volume • Find the area of plane figures • Represent 3-D figures • Measure with standard and metric units • 6. Perform basic operations with whole numbers, fractions, and decimals
Discussion Items	**Materials**
• What do we need to know to solve these problems? • How do you calculate the volume of an object? • Can you defend your reasoning in your solution to this problem?	• Information about barn, hay bales • Diagrams of barn/ hay bales for handouts or transparencies

Lesson Procedures

Ask the students what they know about hay and barns.

Share with students the information about the barn and hay bales.

Give students activity sheets to complete with a partner or in small groups.
- Keep the students engaged by periodically asking their group to reflect on the discussion questions listed above.

Assessment

Use the following rubric to assess student work.

Points	Appropriate Operations and Tools	Written and Oral Explanation
4	*Profound:* a deep, insightful, and reasonable choice of mathematical operations and tools	*Sophisticated:* an exceptionally clear and logical explanation of the correct strategies and solution for the problem
3	*Revealing:* a reasonable choice of mathematical operations and tools	*In Depth:* a logical explanation of the correct strategies and solution of the problem
2	*Able:* an immature choice of mathematical operations and tools	*Developed:* an explanation of the correct strategies, but the solution of the problem was not correct
1	*Apprentice:* an unreasonable choice of mathematical operations and tools	*Novice:* explanation was attempted, but the strategies and solution are not correct

Your Ideas/Comments/Notes for Next Year

About the Author

Sherry Jones completed her doctorate in 2008 and is an Associate Professor of Business Education at Glenville State College in Glenville, WV. She has taught math-oriented courses in the business department at Glenville for 18 years. In 2005, Sherry was awarded the Curtis Elam Professor of Teaching Excellence Award. Prior to teaching at Glenville State, Sherry taught upper level mathematics courses at Gilmer County High School in Glenville, WV, for seven years. She was recognized as a Gilmer County Teacher of the Year for her efforts in this position.

Sherry has lived in Gilmer County all of her life. She and her husband, David, raised two sons, Chris and Casey Jones. Chris and his wife, Amy, have given Sherry what she describes as great joy in her life: two grandsons, Orrin and Elijah Jones.

Sherry can be reached at Sherry.Jones@glenville.edu.

MATHEMATICS IN RURAL APPALACHIA

Student Activity Sheet

Names:_____

Work and explanation on <u>each</u> of the above questions will be assessed by the following rubric. Points for question number <u>two</u> will be *doubled* as this is the main question to be answered in this lesson.

Points	Appropriate Operations and Tools	Written and Oral Explanation
4	*Profound:* a deep, insightful, and reasonable choice of mathematical operations and tools	*Sophisticated:* an exceptionally clear and logical explanation of the correct strategies and solution for the problem
3	*Revealing:* a reasonable choice of mathematical operations and tools	*In Depth:* a logical explanation of the correct strategies and solution of the problem
2	*Able:* an immature choice of mathematical operations and tools	*Developed:* an explanation of the correct strategies, but the solution of the problem was not correct
1	*Apprentice:* an unreasonable choice of mathematical operations and tools	*Novice:* explanation was attempted, but the strategies and solution are not correct

Using all of the given information, answer the following questions and explain orally and in writing how you arrived at your answer.

1. Estimate how many bales of hay can be stored in the hayloft.
2. What is the actual maximum number of bales of hay that can be stored in the hayloft?

Additional Related Questions

3. How much twine will be around the number of bales of hay in your answer to number 1? Give your answer in feet.
4. What geometric figures and/or relationships do you see in this problem?
5. Suppose the floor of the loft will hold no more than 5 tons of weight. Will this change your answer to number 1?
6. If it takes 2 ½ bales of hay per day to feed Farmer Jones' animals during the winter months, how long will the bales of hay in the loft last? (Use the number of bales from your answer to number 1 to help answer this question.)
7. How many cows and horses do you think Farmer Jones has if he uses 2 ½ bales of hay per day to feed them during the winter months?
8. How many times will Farmer Jones have to replenish the hay in the loft during the winter?

Lesson 10

Estimating the Cost of a New Barn Roof

By: Courtenay G. Mayes

Introduction

In this lesson, students were presented with a problem in a real world context and asked to solve it in ways that makes sense to them. Presenting a problem that is relevant to students both motivates and challenges them and also requires them to integrate preexisting and new knowledge, skills, and procedures. From the context of building a barn roof and calculating cost, flows important mathematical ideas including: Pythagorean relationships, slope as it relates to linear functions, triangle properties, and surface area. The goal is for students to see the interconnectedness of the geometrical and algebraic ideas in ways that make sense to them and promote communication, connections, and reasoning. In this lesson, students become autonomous learners as recommended by NCTM's Learning Principle. They are asked to use a variety of processes to complete this lesson including: representing, discussing, analyzing, researching, and writing about mathematical ideas.

Additionally, this lesson can be modified for students from different geographical backgrounds. For example, urban students can calculate the roof of any commercial structure verses a rural barn roof. Urban students will find that the same basic mathematical principles apply.

NCTM Content Standards	NCTM Process Standards
Number and Operations: judge the reasonableness of numerical computations and their results Algebra: Approximate and interpret rates of change from graphical and numerical data. Geometry: Use Cartesian coordinates to analyze geometric situations; draw and construct representations of two– and three dimensional objects Measurement: Understand and use formulas for the area and surface area of geometric figures. Data Analysis and Probability: N/A www.nctm.org	Problem-Solving: Solve problems that arise in mathematics and other contexts. Reasoning & Proof: Make and investigate mathematical conjectures. Communication: Communicate their mathematical thinking coherently and clearly to peers, teachers, and others Connections: Recognize and apply mathematics in contexts outside of mathematics Representations: Use representations to model and interpret physical, social, and mathematical phenomena www.nctm.org
Kentucky State Standards **11th grade assessment**	**Your State Standards** **(fill in here)**
Students use problem-solving processes to develop solutions to relatively complex problems. Students will apply ratios, percents, and proportional reasoning to solve real-world problems (e.g., those involving slope and rate, percent of increase and decrease) and will explain how slope determines a rate of change in linear functions representing real-world problems. Students will determine the surface area and volume of right rectangular prisms, pyramids, cylinders, cones, and spheres in real-world problems. http://www.education.ky.gov/KDE/Instructional+Resources/High+School/Mathematics/default.htm	

MATHEMATICS IN RURAL APPALACHIA

Instructional Plan

Objectives	Prior Knowledge
<ul><li>Students will use Pythagorean relationships to solve problems in real-world situations.</li><li>Students will apply to real-world situations ratio measures including slope.</li><li>Students will calculate surface area.</li></ul>	The students are able to:<ul><li>Define and calculate surface area</li><li>Perform basic operations with whole numbers, fractions, and decimals</li><li>Find the area of plane figures</li><li>Represent 3-D figures</li></ul>
Discussion Items	**Materials**
<ul><li>How many of you have seen a barn roof?</li><li>Are there different types of barn roofs?</li><li>If so, how are they different?</li><li>Why might they be different?</li></ul>	<ul><li>Measuring tools</li><li>Bags of various sizes</li><li>Precise wording of the regulations on the size bags that can be used</li></ul>

Lesson Procedures

Background Information

For roof estimation purposes, it is necessary to estimate the surface area of roofing.

> You can determine the slope of a roof given the base the roof sits on and the desired roof pitch (The pitch is determined by the desired roof attributes and its application).
>
> A roofing square is 10' x 10' section of the roof surface area

Slope is rise over run. Roofing terminology equates distances and slopes on a 12' basis.

> **Example:** 4:12 pitch = a cross sectional rise of 1 foot per 3 feet of linear length

The roof overhang is how much the roof offsets past the base of the building it rests on.

Problem: Determine the surface area and linear footages of materials used to build a roof on a 22' x 48' structure with a 1-foot overhang.

Conceptual understanding:
Calculation of the pitch and both how and why it was determined
Calculation of the exact angle of the roof from the footprint of the building
Calculation of the Materials needed
Calculation of the surface area then the pricing to install the materials
Overall presentation
Conversion of surface area to roofing squares.
Extra prices like vents and gutters

If one looks at the given structure footprint from the top down, this is the area that the roof must cover and protect. The visualization is an empty shell with four walls sticking straight up towards you. The roof is to be set on top of this footprint.

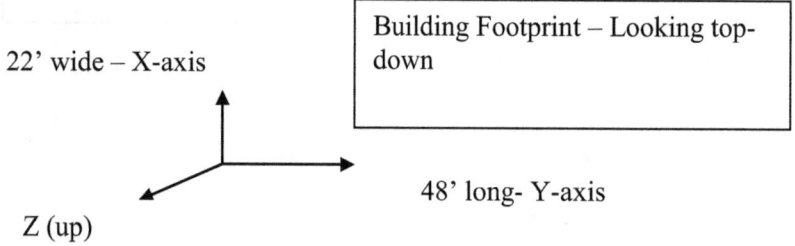

Second, visualize looking at the structure from the side. When looking from a side-view, the roof is visualized as triangles sitting on top of the given structure base wall. That would be similar to standing on the ground and looking at the side of the building with no roof on it. The X-axis would be horizontal when looking at the building from the side.

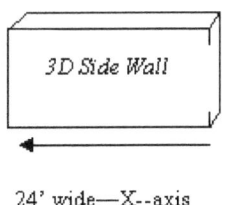

24' wide—X--axis

On top of this sidewall, picture the roof sitting on the wall like a hat. The roof is an obtuse triangle. This obtuse triangle is what the pitch of a roof determines. The pitch of a roof is another way of saying the slope of a roof. The overhang is the distance from the wall to the lower corner of the roof. A 1-foot roof overhang on a 22-foot wide base would mean that the roof base would actually be 24 feet along the X-axis.

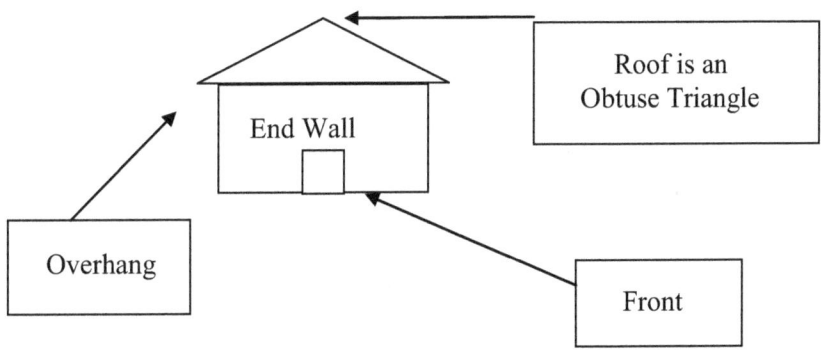

How much does it cost to build a roof?

Scenario:

You have just been awarded a new job with a national roofing-contracting firm. Your job is to design a roof for a client. Your client is a very successful horse owner in Kentucky. Your client needs your help and expertise. Your project is to price the roof on his 5 stables (barns) in rural Kentucky. He requests a 1-foot overhang.

Another building contractor has already built the foundation and provided framework of the building. The deadline for pricing and building is Christmas, so he can present his wife with a new barn. A top view sketch is below.

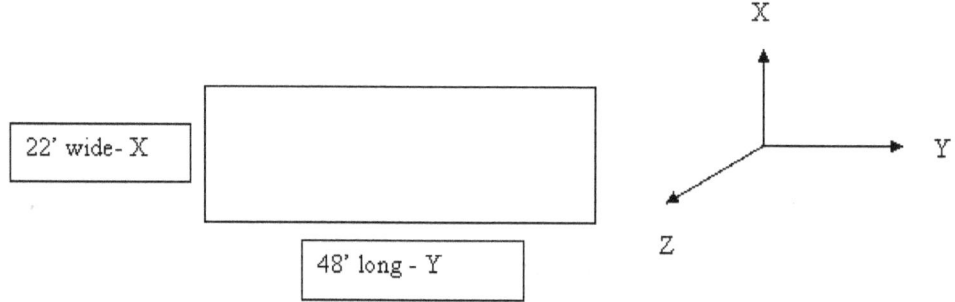

Considerations:

Twelve-foot lengths for roofing materials are often standard for simplicity. In the South, heat builds up and rises to the interior of your roof. You must consider heat rises in your design. In the North, snow can build up on top of your roof and make it collapse unless you find a way to keep this snow from accumulating. You must also consider what you plan to store in the structure to determine what is critical to its design.

Clue: Animals need ventilation in the summer and heat in the winter. A roofing square is a common term roofers use to estimate prices. Material Prices can be obtained from local hardware stores or via the Internet.

Assessment

The customer would like to see the following in your proposal.
- Pan (Goal)
- Design (Calculation using linear functions)
- Cost (Extrapolation)

Remember: Points will be awarded for diagrams, and explanation of calculations. Roof pitch is the slope (rise over run)

Your Ideas/Comments/Notes for Next Year

About the Author

Courtenay Mayes is currently teaching at the Great Oaks Institute of Technology and Career Development in Cincinnati, Ohio. In addition to teaching 11th and 12th graders mathematics, she is occupied with writing her dissertation and trying to coach her teenage daughter through the drama of high school.

MATHEMATICS IN RURAL APPALACHIA

Lesson 11

Round Barns

By: Johnny Belcher

Introduction

Agriculture has always played a large part in the history of West Virginia. Round barns, though never quite as common as the standard square or rectangular shaped barns, were seen as a time and money saver. The selling point for the round barn design was the large loft space, which allowed farmers to store hay in the barn, eliminating the cost of building and maintaining outbuildings for hay storage. This also reduced the farmer's work since they did not need to haul the hay to the barn to feed the livestock.

The Hamilton Round Barn, built in 1912 by Amos C. Hamilton as a dairy barn, also incorporates a technology more commonly associated with the Pennsylvania bank barns. It has two main entrances, one on the lower level for livestock and another, accessible by ramp, to the second level where the farmer would store equipment and silage.

NCTM Content Standards	NCTM Process Standards
Number and Operations: Compute fluently and make reasonable estimates. Algebra: Use mathematical models to represent and understand quantitative relationships. Geometry: Use visualization, spatial reasoning, and geometric modeling to solve problems. Measurement: Understand measurable attributes of objects and the units, systems, and processes of measurement; and Apply appropriate techniques, tools, and formulas to determine measurements. Data Analysis and Probability: N/A www.nctm.org	Problem-Solving: Solve problems that arise in mathematics and other contexts. Reasoning & Proof: Make and investigate mathematical conjectures. Communication: Communicate their mathematical thinking coherently and clearly to peers, teachers, and others Connections: Recognize and apply mathematics in contexts outside of mathematics Representations: Use representations to model and interpret physical, social, and mathematical phenomena www.nctm.org
West Virginia State Standards **6th grade** Solve problems, in context, involving operations on whole numbers, fractions, and decimals. Investigate and model volume and surface area. Apply formulas to determine perimeter, circumference and/or area of plane figures. http://wvde.state.wv.us/csos/	**Your State/District Standards** **(fill in here)**

Instructional Plan

Objectives	Prior Knowledge
Students successfully completing the lesson will be able to: - Explain the need for precision of mathematical statements; - Critically analyze the meanings of a statement involving measurements; - Determine the volume of a rectangular prism; - Estimate the volume of various three dimensional objects; - Physically measure some dimensions and calculate other measurements on various three - dimensional objects; and - Provide logical reasoning for making a decision involving measurements.	The students are able to: - Define and calculate surface area - Define and calculate volume - Find the area of plane figures - Represent 3-D figures - Measure with standard and metric units. - 6. Perform basic operations with whole numbers, fractions, and decimals.
Discussion Questions - Would the way hay is stored today, as compared to 1912, make the round barn more attractive or less attractive? Justify your answer mathematically. - What other shapes might barns be built in? Compare at least two other shapes to a square, rectangle, and circle supporting claims of comparison mathematically.	**Materials** - Diagrams of round barn - Models of objects - Calculators

Lesson Procedures

Activities

There are two statements that I would like for you to justify.

Statement 1: "Round barns, though never quite as common as the standard square or rectangular shaped barns, were seen as a time and money saver."

Statement 2: "The selling point for the round barn design was the large loft space, which allowed farmers to store hay in the barn."

Task 1: Mathematically support or contradict the claim made by statements 1 and 2 in the brochure. (Note: Hay could have been stored in square and rectangular barn lofts as well. So, statement 2 implies that the loft space of the round barn is more efficient.)

Follow-up question: Would the way hay is stored today, as compared to 1912, make the round barn more attractive or less attractive? Justify your answer mathematically.

Task 2: What other shapes might barns be built in? Compare at least two other shapes to a square, rectangle, and circle supporting claims of comparison mathematically.

Extension

A great synthesis project would be to have students design a barn from the ground up where its final use drives the design. Students could be given flexibility in choosing a use for the barn. Cost verses utility and strong mathematical justifications would be the aim of distinguished work. Culminating work could be presented with a schematic or even a scale model along with a presentation.

This is a lesson that allows entry points at multiple levels accommodating for diversity. In particular, the extension allows room for the most advanced student to explore. The mathematics involved can be hands-on, numerical exploration, or all the way to using calculus to maximize volume. The language and presentation of the prompt can be altered to fit all grade levels. Rural students from agricultural communities can relate to this topic, and could be intrigued by the thought of designing a more efficient model of the ordinary barn. This is a topic that can pull students in close to the mathematics and promote a personal ownership in justifications and designs. Student presentations can serve the class and the student best when personal ownership is strongly felt by the presenter.

Your Ideas/Comments/Notes for Next Year

About the Author

Johnny is currently teaching grades 9-12 at Pikeville High School located in Pikeville, the county seat of Pike County. He lives about 25 miles south east of Pikeville close to the Kentucky, Virginia border deep in the Appalachian Mountains, in Dorton, KY where he was raised. He teaches Algebra II, PreCalculus, Calculus I, AP Calculus, and Senior Physics and his wife, Mary L., currently teaches 7th grade science at Pikeville Middle School. They have a daughter, Sarah Elizabeth who is three years old.

Johnny holds a BS in Math/Physical Science in Secondary Education with a MS in Secondary School Counseling, from Alice Lloyd College, a small private work-study college in Pippa Passes, Kentucky with a strong sense for instilling purpose in their students. His MS is from Moorehead State University in Morehead, Ky. Johnny's grandmother was an Appalachian teacher of 42 years who began teaching at a one room school house. Both his father and mother are now retired educators, his father served at Dorton High School, Virgie High School and finally Shelby Valley High School as a Guidance counselor and his mother was a librarian at Robinson Creek Elementary all schools in Pike County.

Johnny has currently found great purpose in working towards his Doctorate Degree in Mathematics Education through the ACCLAIM Program and look forward to the service he can be to his home region and Appalachian communities like it.

Lesson 12

Morgantown Lock and Dam

By: Ron Smith

Picture courtesy of the U.S. Army Corps of Engineers

Introduction

This lesson is designed around the Morgantown Lock and Dam on the Monongahela River. The purpose of this lesson is to learn how slope can be used to describe real events and physical formations while at the same time creating a better appreciation for the importance of the lock and dam to boat traffic on the river. The students will be guided to discover the idea of slope by examining how various aspects of the lock and dam change. This lesson can be used to combine geography, history, science and mathematics together. It takes students away from the idea that mathematics is used to calculate money and commerce and shows them that mathematics is woven throughout science and history. The teacher's role is to ensure that these connections are obvious to the students.

This course will be designed to be used in the 5th through 7th grades to begin to teach the idea of slope to middle school and upper elementary school age children. this approach could also be used in the 8th and 9th grade Algebra I classes to help students see real world applications for slope.

NCTM Content Standards	NCTM Process Standards
Number and Operations: Compute fluently and make reasonable estimates. Algebra: Use mathematical models to represent and understand quantitative relationships. Geometry: Use visualization, spatial reasoning, and geometric modeling to solve problems. Measurement: Understand measurable attributes of objects and the units, systems, and processes of measurement; and Apply appropriate techniques, tools, and formulas to determine measurements. Data Analysis and Probability: Data Analysis and Probability: Find, use, and interpret measures of center and spread, including mean and interquartile range; discuss and understand the correspondence between data sets and their graphical representations, especially histograms, stem-and-leaf plots, box plots, and scatter plots www.nctm.org	Problem-Solving: Solve problems that arise in mathematics and other contexts. Reasoning & Proof: Make and investigate mathematical conjectures. Communication: Communicate their mathematical thinking coherently and clearly to peers, teachers, and others Connections: Recognize and apply mathematics in contexts outside of mathematics Representations: Use representations to model and interpret physical, social, and mathematical phenomena www.nctm.org
West Virginia State Standards **5th-9th grade** Locate and plot points within the four quadrants. Determine the slope of a line given two-points or slope/y-intercept equation ($y=mx+b$). Determine the slope of a line from its graphical representation. http://wvde.state.wv.us/csos/	**Your State Standards** **(fill in here)**

Instructional Plan

Objectives	Prior Knowledge
<ul><li>Students will explore relationships between data and the graphs of lines, especially paying attention to the idea of slope.</li><li>Students will use the coordinate plane, tables, words, and symbolic rules to examine the properties of slopes and lines.</li><li>Students will formulate questions and collect data about different characteristics of the river, lock, dam and commerce.</li><li>Students will formulate conjectures based on the data that they have collected.</li><li>Students will build new mathematical knowledge through their data collection and analysis of data.</li></ul>	<ul><li>The students will need to have an idea of how to compare data.</li><li>No prior knowledge of slope is necessary for the students to do the material in this lesson - although this lesson would be a good supplement for children who have already learned slope, but need to get some idea of where it is used outside of the Cartesian plane.</li></ul>
Discussion Items	**Materials**
<ul><li>What is the slope?</li><li>Can you estimate how many recreational boats will use locks in 5 years?</li><li>What does that mean of "run over the rise"?</li></ul>	<ul><li>Internet & computer</li><li>Map of the surrounding area</li><li>Data about the commerce on the river</li></ul>

Lesson Procedure

Lesson Content

To begin the lesson, the teacher needs to have the students research the history of the locks and dams on the Monongahela River. The students would then be guided to find topics about the dam, locks, river and commerce that change or have changed in the past (in a fairly linear manner). The students can come up with ideas that could include the depth of the river, the amount of cargo that has passed through the locks, the amount of recreational boat traffic use, the depth of the water in the lock while the level is being changed, etc. While some of these may not be linear in nature, most can be described with a linear equation. Students can then find ways to represent the data, which may be with a coordinate graph, but is not limited to a coordinate graph. The students can make conjectures about how their results can be used to estimate the various aspects that they have examined. For instance, students could use their results for the number of recreational boats that use the locks to estimate how many recreational boats will use the locks in 5, 10 or 15 years. The teacher will need to caution the students that these are only estimates, and that there may be other factors involved that would make these assumptions not valid.

The teacher could use the data on the amount of commercial traffic to have a discussion about the reduction of coal production in the area. This could lead to a discussion about how attempts to clean the environment have caused the loss of jobs and created an economic downturn in the local economy. The students could use the depth of the river at each end of the pool that is created by the Morgantown dam, to estimate the amount of drop in height of the river bed in the pool.

In order to tie this into history, the students could be asked to write about how the dams on the river helped towns, like Morgantown, to grow economically. The students could be asked what they feel Morgantown would be like if the dam had not been built at all. To make connections to science, the students could research and write about how dams impact river systems. The students could discuss the benefits and problems that come with dams.

To make connections to slope in the Algebra class, the teacher could have students plot their data on a graph and then assign values for the points. The students could then try to determine how to come up with an equation to describe the slope. Students may come up with the slope as the "run over the rise" instead of the old favorite of "rise over the run". The students would then need a teacher to explain that the method that has been accepted by mathematics is the vertical change divided by the horizontal change. This is a good opportunity for students to see how mathematical ideas may not always be the same, but generally one idea becomes accepted and is used to allow consistency.

Possible Additional Activities:

- Students may not be familiar with how the lock and dam function.
- The students could build a lock and dam model in order to describe how the system works. This could very easily tie into science and possibly art.
- The students could take a field trip to the lock and dam to see the equipment in use.
- A follow up session should include the concept of intercept. The students can use the same data to interpret where the values are at the time that they define as the starting point. This would provide students with a very good example that the initial point of a data set does not have to be 0.

Your Ideas/Comments/Notes for Next Year

About the Author

Ron Smith is a math specialist in Arkansas. He is located at Harding University and works with 17 rural school districts that are located in north-central Arkansas. Ron taught high school and junior high school mathematics for 10 years in Turkey, Arkansas, and Ohio. He is currently working on a doctorate in mathematics education through the ACCLAIM program.

Ron can be contacted at rgsmith@harding.edu.

Lesson 13

The Average Elevation of Decker's Creek Trail

By: Mike Ratliff

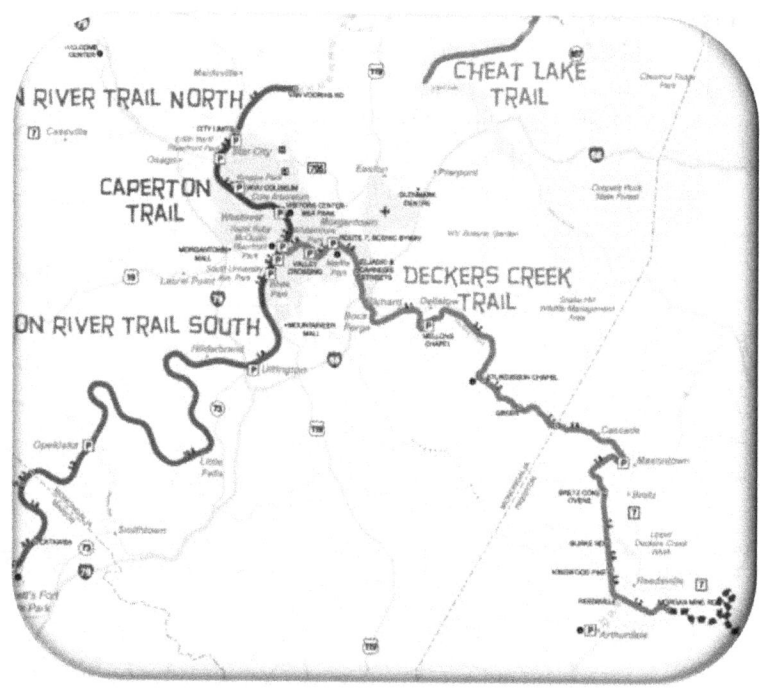

Introduction

In the packet that I received earlier this year from Dr. Mayes regarding West Virginia University and the Morgantown area, a "Rail-Trails of West Virginia" brochure was included. After reading the packet, I decided that I would spend a portion of the July 4th week riding my bicycle on a few of the rail-trails in the Morgantown area. After riding all of the trails in the Morgantown city limits which have a rating of *Easy*, I decided to spend an afternoon conquering the Decker's Creek Trail which has a rating of *More Difficult*. The reason for the increased difficulty is the trail begins in Morgantown and ends near Reedsville, approximately an 18-mile ride (one-way) and an elevation increase of 1,800 feet.

When Kevin Kenady (another cohort member) and I planned the trip, the trail's elevation change created a brief mathematical discussion – especially, the "Decker's Creek Elevation Profile" graph in the brochure. Kevin saw a "slopes" problem, and I saw an "average elevation" problem.

NCTM Content Standards	NCTM Process Standards
Number and Operations: understand numbers, ways of representing numbers, relationships among numbers, and number systems Algebra: analyze change in various contexts Geometry: n/a Measurement: understand measurable attributes of objects and the units, systems, and processes of measurement Data Analysis and Probability: develop and evaluate inferences and predictions that are based on data. www.nctm.org	Problem Solving: solve problems that arise in mathematics and other contexts. Reasoning and Proof: make and investigate mathematical conjectures. Communication: communicate their mathematical thinking coherently and clearly to peers, teachers and others. Connections: recognize and apply mathematics in contexts outside of mathematics Representations: use representations to model and interpret physical, social, and mathematical phenomena. www.nctm.org
INTASC Standards Principle #1: Teachers responsible for mathematics instruction at any level understand the key concepts and procedures of mathematics and have a broad understanding of the K-12 mathematics curriculum. They approach mathematics and the learning of mathematics as more than procedural knowledge. They understand the structures within the discipline, the past and the future of mathematics, and the interaction between technology and the discipline. http://www.ccsso.org/content/pdfs/INTASCMathStandards.pdf	**Your State Standards** **(fill in here)**

Instructional Plan

Objectives	Prior Knowledge
Students successfully completing the lesson will be able to: • Recognize the aspects of a place based lesson; • Be able to align their place based lesson with the NCTM standards; • Recognize and utilize the Principles in the Principles and Standards of School Mathematics	• Familiarity with NCTM's *Principles and Standards for School Mathematics*. • Minimum of 3 years high school mathematics
Key Questions Embedded within the lesson	**Materials** • Bike trail brochure or data points www.montrails.org

Lesson Procedures

What is the Average Elevation of the Decker's Creek Trail?

Description of instructional sequence for the example place-based mathematics lesson: Before addressing the mathematics in this lesson related to rail-trails, a description/history of rail-trails will be provided. The information regarding the description/history of rail-trails that follows will be shared with the class. From the Mon River Trails Conservancy website (www.montrails.org):

Rail-trails are multi-purpose public paths created from former railroad corridors. Flat or following a gentle grade, they traverse urban, suburban and rural America. Ideal for many uses, such as bicycling, walking, in-line skating, cross-country skiing, plus equestrian and wheelchair use, rail-trails are extremely popular as recreation and transportation corridors.

Since the 1960s, almost 11,000 miles of rail-trails have been created across the country. Rail-trails also serve as wildlife conservation corridors, linking isolated parks and creating greenways through developed areas, and as a means of preserving historic landmarks. They often stimulate local economies by increasing tourism and promoting local business.

The abandoned rail corridor that is now the Mon River / Caperton / Decker's Creek Trail System, was purchased from CSX Railroad and is now held in a lease with the WV State Rail Authority. To date, over $2 million in Transportation Enhancement funds have been invested in the purchase and construction of this trail system. These funds have been further augmented by business and community donations.

In the "Rail-Trails of West Virginia" brochure, a map showing the Decker's Creek Trail and a graphical representation of elevation changes are provided. The graphical representation of the elevation changes prompted many questions/problems including the following: What is the average elevation of the Decker's Creek Trail? Students will have that question/problem posed to them, a few minutes to work in groups on the problem, and then an opportunity to present their ideas for finding the answer. (I anticipate that most groups will suggest averaging the minimum and maximum elevations from the graph. My hope is that there will be at least one group who approaches the problem from a sampling perspective. If not, I will guide the class with probing questions.)

Approaching the problem from a sampling perspective presents other obstacles, namely, 1) What points (and how many) on the trail (which is continuous) are included in the sample? and 2) How is the elevation at a point determined? The answer to the second question is elevation can be determined at a point (location on the trail) using a GPS – Global Positioning System. A GPS is an electronic device that allows one to determine exact location (longitude/latitude), elevation, and time most anywhere on Earth. (A GPS is a satellite navigation system first used by the US military and now available at most stores who have sporting goods. The cost of a GPS at discount stores range from $100 to $250.) Now, the first question has multiple answers. My intention is to promote discussion regarding the trail being continuous with an infinite number of points.

One way of approaching the problem is to define the population as being the set of points at each 1/10 of a mile from the trail's beginning to end. When students do the arithmetic, they'll see that this definition provides is at least 180 data values. (The actual length of the trail is between 18 and 19 miles; some brochures indicate 18 miles and others 19 miles.) So, if one programs a GPS to record elevation data values each 1/10 mile and attaches it to a bicycle, the data collection can be completed fairly easily. (According to the Pathfinder bicycle shop in Morgantown, a mounting bracket for a GPS can be purchased and easily mounted on a bicycle. I will have already collected the data on a previous ride of the trail.) Once the data is presented to students, most (if not all) will average the data values producing a number representing the average elevation for the trail. For the defined population, this value is the population mean; I'll then proceed to defining and discussing random sampling. (Other samples such as convenience and voluntary will be discussed briefly.)

After a brief discussion of random sampling, students will practice the technique by selecting a random sample of size 10 from the defined population. (Random numbers using a TI-84 calculator or a spreadsheet will be used to generate samples.) Students will then compute the sample mean for their sample and compare it with the population mean. To demonstrate the relationship between sample and population means, the class will construct a dot plot (the TI-84 can be used to do this by modifying the scatter plot function). This relationship is difficult to see with 30 or less sample means, so students will be asked to generate three or four sample means for plotting. After the dot plot is completed, the population mean will be labeled and the question "How would you describe this dot plot?" will be posed. (If the plot is not normal (which is unlikely), more sample means can be computed until the normal curve appears.) A brief discussion about sample and population means (i.e., the distribution of sample means about a population mean) will follow the display of the dot plot.

Following this discussion, I will pose the question, "What will happen if we increase our sample size to 25? Will the plot be the same or different?" I'll solicit responses from the class.

The graphical representation of the elevation changes prompted many questions/problems including the following: What is the average elevation of the Decker's Creek Trail? Students will have that question/problem posed to them, a few minutes to work in groups on the problem, and then an opportunity to present their ideas for finding the answer. (I anticipate that most groups will suggest averaging the minimum and maximum elevations from the graph. My hope is that there will be at least one group who approaches the problem from a sampling perspective. If not, I will guide the class with probing questions.)

Assessment

The purpose of the previous lesson is to expose students in a secondary mathematics methods course to a place-based lesson. Their assignment is as follows:

1) Identify the standards and expectations from PSSM addressed in the lesson. Be sure to justify your selections.

2) Review "The Technology Principle" (PSSM) and reflect on today's usage of technology. Be prepared to discuss this lesson in the context of this principle.

3) Develop a place-based lesson with your "home" town/community/county as *place*.

Your Ideas/Comments/Notes for Next Year

About the Author

Michael Ratliff, ratliffm@lindsey.edu, teaches mathematics at Lindsey Wilson College (Columbia, Kentucky). He is pursuing a doctorate in mathematics education through the University of Tennessee and ACCLAIM, a center for learning and teaching, sponsored by the National Science Foundation. Ratliff is interested in the mathematical content knowledge needed for teaching geometry at the middle and high school levels. He is also interested in lesson study groups and place-based mathematics education in rural settings.

Lesson 14

Are We Really Mountaineers?

By: Nicolyn Smith

Introduction

This lesson was inspired by a visit to Cooper's Rock State Forest on July 21, 2005. The lesson is directed toward mathematics but also incorporates geography, writing, and history.

The Appalachian Mountain Range that surrounds Morgantown, West Virginia is made up of several smaller ranges. How do these ranges compare with other mountains in North American and the world?

National Standards	**Process Standards**
Number and Operations: Understand numbers, ways of representing numbers, relationships among numbers, and number systems. Algebra: n/a Geometry: n/a Measurement: Understand measurable attributes of objects and the units, systems, and processes of measurement; Data Analysis and Probability: n/a www.nctm.org	Problem-Solving: Solve problems that arise in mathematics and other contexts. Reasoning & Proof: Make and investigate mathematical conjectures. Communication: Communicate their mathematical thinking coherently and clearly to peers, teachers, and others Connections: Recognize and apply mathematics in contexts outside of mathematics Representations: Use representations to model and interpret physical, social, and mathematical phenomena www.nctm.org
West Virginia State Standards **6th grade** Solve problems, in context, involving operations on whole numbers, fractions, and decimals. Demonstrate understanding of measurable attributes of objects and the units, systems, and processes of measurement http://wvde.state.wv.us/csos/	**Your State Standards** **(fill in here)**

MATHEMATICS IN RURAL APPALACHIA

Instructional Plan

Objectives	Prior Knowledge
Students successfully completing the lesson will be able to: • Use the library or internet to locate information • Round large numbers • Compare data values using ratios expressed as fractions and percents • Work with a partner to organize and communicate their findings	• Students have previously worked with ratios and percents using smaller values. • This lesson is an opportunity to work with real data that involves larger values.
Discussion Items	**Materials**
• Is the median always a value in the data set? • When is the median not a member of the data set?	Helpful websites: • www.peakware.com • www.peakbagger.com • http://freespace.virgin.net/john.cletheroe/usa_can/usa/appalach.htm Other Resources: • *A Walk in the Woods* by Bill Bryson

MATHEMATICS IN RURAL APPALACHIA

Lesson Procedures

Have students conduct the research necessary to complete the table on the accompanying sheet. Students may work with a partner and use the internet or library to research mountain ranges. If time is limited, the instructor may provide the web sites or the information rather than allowing the students to conduct the search. The accompanying sheet for collecting data lists the measurements in meters. The instructor may want the students to record the measurements in feet also to compare the two measurement systems

Additional lessons that may use the same theme:

1. It is your job to estimate the height of a mountain. How will you do it?
   - This could lead into a discussion of right triangles and cones or researching other means that are used to determine the height of mountains.
   - This may serve as a lesson on measurement and estimating. Students will work in pairs on the investigation and write their own individual responses to the questions.

2. Investigate the story of Grandma Gatewood who hiked the Appalachian Trail.
    a. What surprised you about her story?
    b. What was the length of her hike?
    c. With your partner estimate how long it would take you to hike the Appalachian Trail and describe how you arrived at the estimate.

Your Ideas/Comments/Notes for Next Year

About the Author

Nicolyn (Nickie) resides in rural Bidwell, Ohio with her husband, Dewey. She has three nephews and two nieces. Nickie earned a Bachelor of Science degree in mathematics and physics along with secondary education from Rio Grande College and a Master of Science degree in mathematics from Ohio University. Nickie is currently an Assistant Professor in the Math and Computer Science Departments at the University of Rio Grande which is locate in the village of Rio Grande in southern Ohio. The institution includes a community college within a four-year university. She has served on a variety of committees at Rio Grande and participates in the University's Masterworks Chorale. Nickie and her husband are active in the River of Life United Methodist Church in the Foothills District of the West Ohio Conference.

Nickie can be reached at nsmith@rio.com
Formatting problems??

MATHEMATICS IN RURAL APPALACHIA

Are we really Mountaineers?

The Appalachian Mountain range that surrounds Morgantown, WV is made up of several smaller ranges. How do these compare with other mountains in North America and the world?

Explain how we can compare the height of the highest peak in the world with the height of the highest peak in the Appalachian Mountains. Include a statement in which you express the ratio as both a fraction and a percent.

Write another piece of information that you discovered during your research that you can share with the class.

Are we mountaineers? Why?

	Name	Mountain Range	Location (state/province)	Actual height in meters	Height rounded to the nearest meter
Tallest mountain in the world					
Tallest mountain in North America					

Names of ranges in the Appalachian Mountains	Location (states/ provinces)	Name of tallest mountain in the range	Actual height of highest mountain in meters	Height rounded to the nearest meter	Write a ratio as a fraction to compare this height to that of the highest peak in the world	Write this ratio as a percent

Lesson 15

The 4-H Club and Farm Animals

By: Nicolyn Smith

Introduction

4-H is an organization for youth that are approximately 8 – 19 years of age. Each member completes one or more projects for the purpose of learning more about the subject area and enters the project for competition at a local county fair. He or she may be chosen to compete at the state level. Many of the projects involve raising animals, and in rural areas many of those are animals commonly found on farms such as beef and dairy cattle, swine, sheep, poultry, and horses. A member need not live on a farm in order to raise what is considered a farm animal. Surprisingly, many hogs and other animals are raised on "non-farms" in rural areas.

Data pertaining to 4-H and animal projects in the state of Ohio can be found at www.ohio4h.org/. The teacher may also need to make arrangements months in advance so that appropriate data for the county and individual projects can be obtained. Suggestions: contact the local extension agent and the fair board and ask what records are maintained by these offices and ask students in the school to collect appropriate data from January through September in the year prior to using the lesson. Contact a local farmer or the local cooperative extension agent or a representative from the state extension office who will address the class on a topic of interest. Ask students in the class or in the school to share their own stories and pictures. This may serve as a sociology lesson to those who are not familiar with farm life or raising animals and as an opportunity for those with experience to share what they have learned.

NCTM Content Standards	NCTM Process Standards
Number and Operations: Work flexibly with fractions decimals and percents to solve problems; understand and use ratios and proportions to represent quantitative relationships; understand the meaning and effects of arithmetic operations with fractions, decimals and percents Algebra: n/a Geometry: n/a Measurement: n/a Data Analysis and Probability: n/a www.nctm.org	Problem Solving: solve problems that arise in mathematics and other contexts Reasoning and Proof: n/a Communication: organize and consolidate their mathematical thinking through communication Connections: recognize and used connections among mathematical ideas; understand how mathematical ideas interconnect and build on one another to produce a coherent whole. Representation: select, apply, and translate among mathematical representations to solve problems www.nctm.org
Ohio State Standards **Grades 5-7** Compare, order and convert among fractions, decimals and percents. Use models and pictures to relate concepts of ratio, proportion and percent. Recognize whether an estimate or an exact solution is appropriate for a given problem situation. http://www.ode.state.oh.us/academic_content_standards/acsmath.asp	**Your State Standards** **(fill in here)**

Instructional Plan

Objectives	Prior Knowledge
Students successfully completing the lesson will be able to: • See the relationship of fractions, decimals, and percents, and their use in comparison and solving problems. • Prior to this lesson, students have worked with rational numbers as fractions, decimals, and percents and with ratios and proportions.	• Fractions • Decimals • Percents • Ratios • Proportions
Key Questions • What groups or clubs do you belong to? • Do you participate in 4-H • Besides a fraction, what other ways can you represent this data • How to you change between representations?	**Materials** • Computer Lab • Internet

MATHEMATICS IN RURAL APPALACHIA

Lesson Procedures

By the time children enter grade 6, many are members of small groups. Divide the students into small groups in the classroom. Ask each group to list the other small groups or organizations of which they are a part or to which they belong. One member of each group will list their groups or organizations on the board. The groups may include a family, a Sunday School class, a sports team, etc. As a class, discuss what has been learned by being a part of these groups. These may include getting along with others, athletic skills, responsibility, etc. Did any of the groups list 4-H? Does anyone know the origin of the name 4-H? What projects have students taken as 4-H members?

The examples below involve swine and feeder calf projects. The questions are appropriate for other market animals such as steers and sheep. Allow students to continue to work in the same small groups.

For the questions, ask students to express the "fraction" in fraction, decimal, and percent form. Gallia County is used here for example only, and the most recent data available is for 2004. The local data will be provided and data for the state is available online. The teacher may choose to have all the groups work on all of the questions or randomly assign questions to groups. Each student in a group will write his name at the top of a sheet of paper and list the names of the other members of his/her group and will keep a neat and accurate record of the work of the group. These sheets and classroom observation by the teacher will serve as an informal assessment. Students may be asked to share their group's results in class. What follows is a worksheet that can be completed for homework and used for assessment.

1) What fraction of all hogs raised as projects in Gallia County are *home grown*? (There is a date by which animals taken as projects must be identified, but they need not be born in the county.)

   During the last ten years, what fraction of the hogs who won grand champion or reserve champion were *home grown*?

   Write a sentence in which you compare these results.

It will be best if a computer with internet access is available for each small group.
Use the web site www.ohio4h.org/ to answer questions like the following:

2) What fraction of the counties have 4-H members who reside on farms only?

   What fraction of the counties have 4-H members who reside on farms or in rural areas only?

   What fraction of the 4-H members in Gallia County reside on a farm?

MATHEMATICS IN RURAL APPALACHIA

Choose a neighboring county. What fraction of the 4-H members there reside on a farm?

Write a sentence that tells which fraction is greater and how much greater.

Consider Franklin County. What fraction of the 4-H members there reside on a farm?

3) What fraction of the state projects are swine projects?

What fraction of the projects in Gallia County are swine projects?

Write a sentence that tells which fraction is greater and how much greater.

4) One online table shows the names of the counties and the number of 4-H members in each county and the number of clubs in the county, and the percent of the county population that is 5-19 years of age.

Does Gallia County or Franklin County in Ohio have a higher rate of youth, 5-19, participating in 4-H? What additional information, if any, is needed to answer the question? Where will you find the information?

Ask each small group to use the online data and write a question that they will ask of another group.

Your Ideas/Comments/Notes for Next Year

About the Author

Nicolyn (Nickie) resides in rural Bidwell, Ohio with her husband, Dewey. She has three nephews and two nieces. Nickie earned a Bachelor of Science degree in mathematics and physics along with secondary education from Rio Grande College and a Master of Science degree in mathematics from Ohio University. Nickie is currently an Assistant Professor in the Math and Computer Science Departments at the University of Rio Grande which is locate in the village of Rio Grande in southern Ohio. The institution includes a community college within a four-year university. She has served on a variety of committees at Rio Grande and participate in the University's Masterworks Chorale. Nickie and her husband are active in the River of Life United Methodist Church in the Foothills District of the West Ohio Conference.

Nickie can be reached at nsmith@rio.com

MATHEMATICS IN RURAL APPALACHIA

Assessment: Homework/Assessment Due _____
Fractions, Decimals, Percents

1) Express the given fraction in decimal form and as a percent.
   Express the given decimal as a fraction in lowest terms and as a percent.
   Express the given percent as a fraction in lowest terms and in decimal form.

   $\frac{4}{5}$

   .375

   5.5%

2) Compare $\frac{1}{3}$ and 30%. Which is larger? Explain how you know.

3) What is 35% of 320 pounds? Show how you arrived at your answer.

4) 25 pounds is 40% of what weight? Show how you arrived at your answer

Feeder Calf Project Animal Inventory

ANIMAL INFORMATION				BEGINNING ANIMAL INVENTORY					CLOSING ANIMAL INVENTORY		
Identification of Animal (include all available information)				Date Obtained (mm/dd/yy)	Purchased (Also, include Name and Address of Breeder or Producer) **	Date & Estimated Weight	Purchase Price or Value at Start of Project	Comparison Price (Market Value)	Date & Estimated Weight	Kept	Sold
Name and ID # (tattoo and/or Tag #)	Description (Breed, color, markings, etc.)	Sex	Birthdate of animal (mm/dd/yy)	Raised (Born)						Value at End of Project	Total Selling Price
	Shorthorn	Male					$	$		$	$
							$	$		$	$

** It is highly recommended that you record the Name and Address of the breeder or producer, from whom you purchased each animal.

Lesson 16

Dent's Run Covered Bridge

By: Debbie Waggoner

Photo courtesy of the West Virginia Division of Tourism

Introduction

The Dent's Run covered bridge is the only remaining covered bridge in Monongalia County, and one of only 17 covered bridges left in West Virginia. The bridge was originally built in 1889 and was restored in 1984. It is still open for traffic but is not in use because there is an adjacent concrete bridge. Covered bridges are usually some type of truss beam bridge with a covering to help protect the truss and prevent extra weight, such as snow, from exerting pressure on the deck.

The primary purpose of covering the deck and trusses was to shield the bridge from snow and rain to keep the wood from decaying and rotting to ensure a longer lifetime for the bridge. Engineers have stated that a covered timber truss has a life expectancy at least three times greater than one that is not covered. Most of America's covered bridges were built between 1825 and 1875. Later concrete and iron began to be used because they were stronger than wood and were not susceptible to fire.

The Dents Run covered bridge was designed with a Kingpost truss design that is 40' long and 12'10" wide. The basic kingpost truss features a center upright post (which gives the truss its name) framed into a triangle formed by the bottom chord and two diagonals. All three timbers are in compression, taking the burden up to the peak of the pyramid. Kingpost trusses are widely used in building construction, but have a serious limitation when used for modern bridges: the heavier the load to be carried, the longer the diagonal timbers must be.

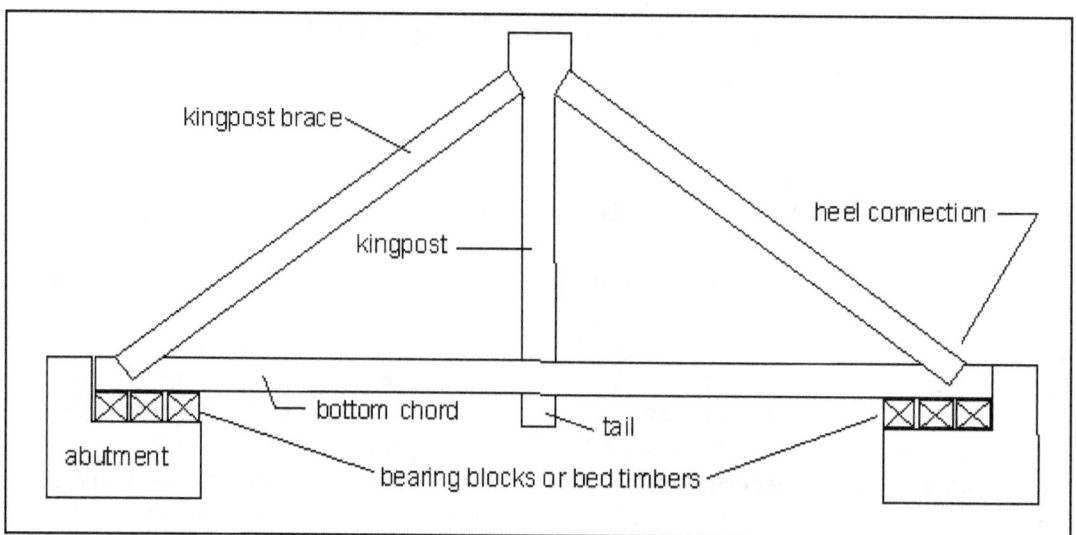

Sample diagrams of kingpost truss design
(Courtesy of the United States Department of Transportation)

NCTM Content Standards	NCTM Process Standards
Number and Operations: Compute fluently and make reasonable estimates. Algebra: N/A Geometry: Use visualization, spatial reasoning, and geometric modeling to solve problems. Measurement: Understand measurable attributes of objects and the units, systems, and processes of measurement; and Apply appropriate techniques, tools, and formulas to determine measurements. Data Analysis and Probability: N/A www.nctm.org	Problem-Solving: Solve problems that arise in mathematics and other contexts. Reasoning & Proof: N/A Communication: N/A Connections: Recognize and apply mathematics in contexts outside of mathematics Representations: N/A www.nctm.org
West Virginia State Standards (6th grade) Demonstrate understanding of numbers, ways of representing numbers, and relationships among numbers and number systems; Solve ratio and proportion problems including scale drawings and similar polygons Apply appropriate techniques, tools and formulas to determine measurements through communication, representation, reasoning and proof, problem solving, and making connections within and beyond the field of mathematics http://wvde.state.wv.us/csos/	**Your State/District Standards (fill in here)**

Instructional Plan

Objectives

- Students will be able to understand and use ratios and proportions to represent quantitative relationships
- Students will be able to understand relationships among similar objects, describe shapes under scaling, and recognize and apply geometric ideas and relationships to areas outside the mathematics classroom
- Students will be able to solve problems involving scale factors, using ratio and proportion

Prior Knowledge

The students are expected to know the knowledge of :
- Ratios
- Proportions
- Scaling
- Similarity

Discussion Items

- How did you come up with the design for your bridge?
- Which geometric shapes did you use in your bridge?
- Why did you use those shapes?
- Where can you find information to help you decide how much it would cost to actually build your bridge?
- Why would your community want to reconstruct a covered bridge opposed to another type of bridge?

Materials

- Graph paper
- Popsicle sticks
- Glue
- Cardboard base
- Ruler

Lesson Procedures

Launch

Engineers first create a blueprint and model of a bridge before they begin construction. Models enable them to test the design of their bridges. Often, engineering companies must compete to win a contract. For their presentations, they explain features of their designs with blueprints and models. Suppose there is a bid out for the reconstruction of a covered bridge that has been damaged in your area. Your group will be an engineering team trying to win the bin for the restoration.

Explore

Each group will design and build a covered truss bridge model for the Dents Run covered bridge which is located here in Monongalia County.
-- First read over the introduction about the Dents Run Covered Bridge Information then create a blueprint (scale drawing) of your bridge design on graph paper. Be sure to draw and label your design to scale using the correct dimensions of 40' long and 12'10" wide as provided on the information sheet.
-- Second, using your blueprint, estimate how many popsicle sticks you will need for the model. When you have your estimate have one person from your group get all the materials.
-- Third, create a scale model of your bridge using popsicle sticks, glue, a cardboard base and a ruler.

Summarize

Plan how you will present your bridge to the large group tomorrow by explaining the mathematics you used and the rationale behind your design.

Assessment

With all the groups together, test each of the bridges for length, height, and strength. Invite a civil engineer to talk to your class about bridges. What other types of bridges exist in your area? Which mathematics and science courses did the engineer take to prepare for a career in engineering? What tools do engineers use to design bridges and other structures?

Additional Related Ideas

Trusses have a high strength to weight ratio and consequently are used in many structures, from bridges, to roof supports, to space stations - the online program, BRIDGE DESIGNER actually allows students design trusses: http://www.jhu.edu/virtlab/bridge/truss.htm

MATHEMATICS IN RURAL APPALACHIA

Your Ideas/Comments/Notes for Next Year

Lesson 17

Rodeo and Statistics Lessons for Middle School Students

By: Jamie Fugitt

Introduction

In the Ozarks Region of southwest Missouri owning one or several horses is a very common occurrence. Many small towns in rural Missouri have a rodeo arena where people gather on a regular basis to practice various rodeo and riding skills and to visit with family and friends. Even in small towns where this is not a regular occurrence, a rodeo is often an annual event at the local fair. Because of these experiences children in the rural communities often are very interested in and familiar with caring for horses, riding horses, and rodeo activities. This serves as the motivation for the theme of the unit which will be described briefly below and two mathematics lessons will be described briefly and then followed by the lesson plans for the two lessons.

The two lessons which follow are part of a larger unit designed to study the local rodeo in Forsyth, Missouri. The unit is interdisciplinary focusing on history, economics, writing, and mathematics. During the unit the students will research the history of rodeos in general and specifically the rodeo held in Forsyth. Students will gather data to use as they study the economic impact of the rodeo on the local economy. Students will write papers and give presentations about these concepts. If possible the students will become involved in providing advertisement for the rodeo and doing volunteer work at the rodeo.

NCTM Content Standards	NCTM Process Standards
Number and Operations: Computes fluently and makes reasonable estimates Algebra: Use mathematical models to represent and understand quantitative relationships Geometry: n/a Measurement: n/a Data Analysis and Probability: Find, use, and interpret measures of center and spread, including mean and inter-quartile range; discuss and understand the correspondence between data sets and their graphical representations, especially histograms, stem-and-leaf plots, box plots, and scatter plots www.nctm.org	Problem-Solving: Solve problems that arise in mathematics and other contexts. Reasoning & Proof: n/a Communication: Communicate their mathematical thinking coherently and clearly to peers, teachers, and others Connections: Recognize and apply mathematics in contexts outside of mathematics Representations: Use representations to model and interpret physical, social, and mathematical phenomena www.nctm.org
Missouri State Standards Find, use, and interpret measures of center, outliers and spread, including range and inter-quartile range Compare different representations of the same data and evaluate how well each representation shows important aspects of the data http://dese.mo.gov/divimprove/curriculum/GLE/MathFinalGLE_3.2.04.pdf	**Your State/District Standards (fill in here)**

MATHEMATICS IN RURAL APPALACHIA

Instructional Plan - Rodeo Lesson Part I

Objectives	Prior Knowledge
Upon successful completion of the lesson the students will be able to: - Determine the median, range, quartiles, and interquartile range of a set of data; - Explain the processes used to determine each of the above measures; - Explain the meaning and use of each of the above measures; - Begin to relate each of the above measures to a box plot	The students are able to: - Arrange numbers recorded to the nearest tenth of a unit in order from smallest to largest - Students also are expected to have some knowledge of stating the median of a set of data, although it is not critical to the lesson.
Discussion Items	**Materials**
- Is the median always a value in the data set? - When is the median not a member of the data set?	- Note cards containing sets of data - Prepared box plot - Calculators - Worksheet 1

Lesson Procedures

Anticipatory Set

If possible have a student talk about the sport of barrel racing and/or bull riding explaining the event and how it is scored. The student should also explain the typical range of scores for the event. If it is not possible to have a student speak to the class, have an adult involved in rodeo speak to the class or present rodeo information from a video or pictures and information found on the web.

Instruction/Modeling and Guided Practice

Activity 1
- Divide students into groups of 3–6 students.
- Give each group a set of note cards, each of which contains one number. These should be made from the data included in the tables at the end of this document.
- Direct each group of students to arrange their set of cards in order from smallest to largest numbers and determine what number is the center of the data. If possible display the numbers by setting them in the chalkboard tray or taping them to the wall.
- Have each group report on how they determined the middle number in the set of data. Ask if someone knows what this middle number is called? (median)
- If the differences of determining the median for an odd number of data and an even number of data have not been adequately discussed, solicit this information from students by asking questions such as: "Is the median always one of the data?" "When is the median not one of the data?" "How do you determine the median when you have an odd number of data?" etc.
- Ask each group to now determine the median for each half of their data using the processes discussed above. Ask if someone knows what the median of the first half of the data is called (Lower Quartile or First Quartile or Q_1) and what the median of the second half of the data is called (Upper Quartile or Third Quartile or Q_3).
- Discuss, asking questions of the students, range and interquartile range of a set of data and the use of these measures.
  To assess understanding of median, quartile 1, quartile 3, range, and interquartile range have students work in groups on questions 1 through 4 from worksheet 1 (page 10).

Activity 2
- Display the box plot titled *Barrel Racing Times for 2002 Forsyth Rodeo*. Have students display the corresponding times below the number line (pages 8 & 9).
- Working in groups students should discuss connections between the box plot and the measures discussed in activity 1 (median, quartile 1, quartile 3, range, and inter-quartile range). Have each group record their findings on a poster which can be displayed for the class.

Closure

Today we have looked at some very useful measures for sets of data. These include quartiles, median, range, and inter-quartile range. We have just begun to see how these measures are then used in box plots. In tomorrow's lesson we will learn more about using these plots including using a computer to construct box plots.

Independent Practice

Complete question 5 from worksheet 1 and be prepared to discuss in class tomorrow.

Your Ideas/Comments/Notes for Next Year

MATHEMATICS IN RURAL APPALACHIA

DATA Sheet 1: Separate for students to use during classroom activity

Barrel Racing Times for the
2003 Forsyth Rodeo

Contestant Number	Time (in seconds)
1	18.2
2	19.5
3	17.6
4	20.4
5	21.6
6	18.0
7	19.6
8	22.4
9	21.8
10	22.5
11	24.6
12	21.5
13	20.2
14	20.4
15	26.0
16	19.3

Barrel Racing Times for the
2004 Forsyth Rodeo

Contestant Number	Time (in seconds)
1	17.2
2	18.9
3	16.9
4	25.4
5	18.6
6	22.7
7	16.9
8	17.4
9	17.9
10	18.0
11	18.2
12	18.3
13	21.4
14	22.6
15	21.0

DATA Sheet 2: Separate for students to use during classroom activity

Bull Riding Scores for the
2003 Forsyth Rodeo

Bull Riding Scores for the
2004 Forsyth Rodeo

Contestant Number	Score
1	76.5
2	88.2
3	79.6
4	77.4
5	89.7
6	90.6
7	72.4
8	68.5
9	70.2
10	72.4
11	81.5
12	80.7
13	88.4
14	90.9
15	91.5
16	89.6
17	69.4

Contestant Number	Score
1	78.4
2	82.1
3	89.3
4	92.8
5	93.4
6	88.6
7	73.7
8	72.5
9	82.3
10	80.5
11	90.3
12	91.6
13	75.4
14	77.2
15	78.0
16	84.5
17	90.0
18	91.5
19	86.4

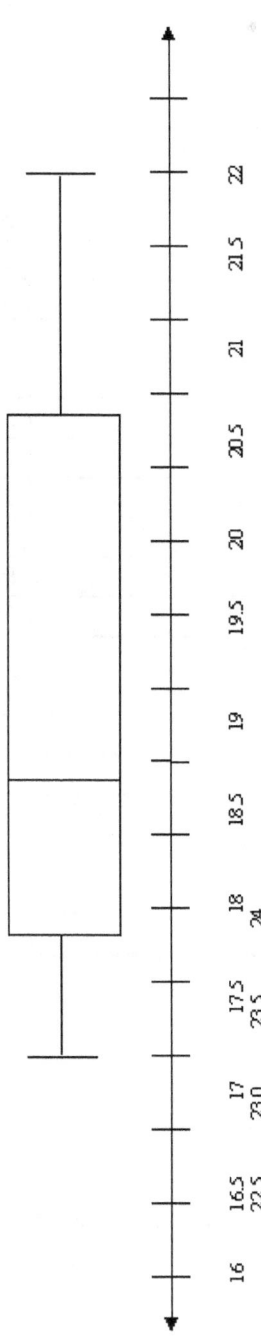

MATHEMATICS IN RURAL APPALACHIA

Barrel Racing Times for Forsyth Rodeo 2002

Contestant Number	Time (in seconds)
1	21.0
2	20.5
3	21.1
4	22.6
5	23.5
6	17.5
7	19.0
8	18.8
9	17.7
10	23.1
11	19.4
12	18.6
13	18.0

NOTE: A large copy of this box plot can be displayed in the front of the room for a large group activity or a small copy can be given to each group. The times from the chart can be cut apart so that the students can place the data below the number line in order to see the connection between the data and the box plot.

MATHEMATICS IN RURAL APPALACHIA

Rodeo Statistics Worksheet 1   Name_____

1. The median score for a set of 15 bull riding scores is 78.6. List some facts about the data that you can conclude as a result of knowing this information.

2. The first quartile for a set of 20 barrel racing times is 17.6 seconds and the third quartile is 22.4 seconds?

    (a) What is the range of the middle 50% of the times?

    (b) How many of the riders had a time of 17.6 seconds or faster?

    (c) What do you know about the times of the slowest one-fourth of the riders?

    (d) Can the median of the data be determined from the information given? Explain.

3. For a set of 30 barrel racing times the difference between the slowest time and the lower quartile is 12.7 seconds while the difference between the third quartile and the fastest time is 3.5 seconds. Explain what could have occurred in the event to account for these times.

4. Construct a set of nine bull riding scores which have a median of 75.4, a lower quartile of 69.3 and an upper quartile of 78.4.

    (a) What are the range and interquartile range for your set of data? Will these be the same for any set of data which satisfies the above information?

    (b) If possible create another set of scores satisfying the above information and having a range of 10 points.

    (c) What is the smallest range a set of scores can have and still satisfy the conditions given in the problem?

5. Seventeen calf-roping times (in seconds) were used to construct the box plot shown below. Record 5 true statements about the data which can be concluded from the graph. Be prepared to share at least one of your statements with the class tomorrow.

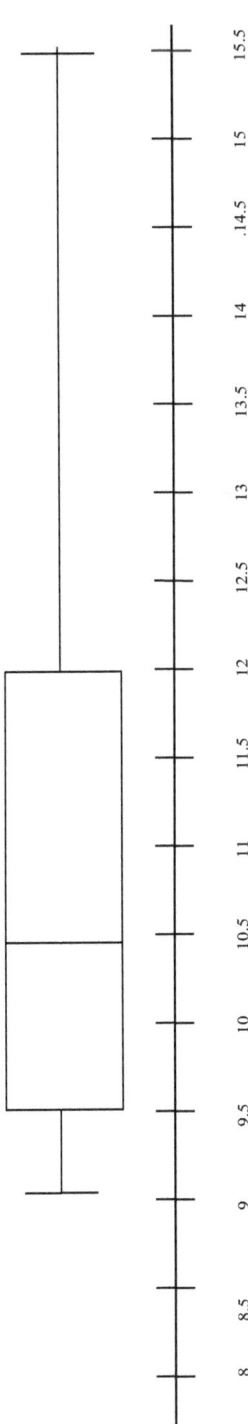

Calf-Roping Times from the 2003 Forsyth Rodeo

MATHEMATICS IN RURAL APPALACHIA

Instructional Plan - Rodeo Lesson Part II

Objectives

Upon successful completion of the lesson the students will be able to:

- Construct a box plot for a set of data;
- Given a box plot for a set of data;
- Determine the range, interquartile range, median, lower quartile, and upper quartile for the set of data;
- Given a box plot, make generalizations about the set of data

Prior Knowledge

In order to successfully complete this lesson, students need to have completed the first part of the lesson so that they have familiarity with range, interquartile range, median and quartiles.

Discussion Items

- Compare your responses to #1 from yesterday.
- How do you construct a box and whisker plot?
- What information do you need?

Materials

- Computers with internet access
- Large box plot
- Data recorded on index cards
- Worksheet 2
- Project Assignment

Lesson Procedures

Anticipatory Set

Display a large copy of the box plot *Barrel Racing Times from Forsyth Rodeo 2002* with the data displayed below the number line (pages 8 & 9). Ask students to pair up and compare their responses to worksheet 1 question #5. Each pair of students should select one response to question #5 and write it on a 4" x 6" index card. The index cards should then be posted next to the box plot. Read and discuss each statement correcting any incorrect statements.

Instruction/Modeling and Guided Practice

Lead students in a discussion about how to construct a box plot allowing students to provide as much of the detail as they can. Record the steps on a poster so that they can remain posted in the classroom. These steps should include the following:

>   Order the data from smallest to largest.
>   Make a number line, being careful to include the smallest and largest datum.
>   Determine the lowest datum, the highest datum, Q_1, Q_3, and the median and place a dot above the number line for each of these numbers.
>   Draw a rectangle from Q_1 to Q_3 divided at the median.
>   Draw a line to connect the lowest datum to Q_1 and Q_3 to the highest datum.

Working in groups, have students construct a box plot and respond to the questions on item 1 on worksheet 2. As the students work the teacher will move from group to group monitoring the progress of the groups.

If equipment is available demonstrate *Box Plot* software from the NCTM website available at http://illuminations.nctm.org/tools/tool_detail.aspx?id=77.

Allow students to construct and print the box plot described in question 2 on worksheet 2.

Independent Practice and Closure

Discuss with students the project, *Display Your Own Data*, which is included on page 15.

About the Author

In addition to completing a doctoral degree through ACCLAIM and being employed full-time at College of the Ozarks, where she serves as Chair of the Division of Mathematical and Natural Sciences, Jamie enjoys spending time with her family. Her husband Jeff is also a doctoral student pursing a degree in religious studies and the social sciences. Their son Jonathan is 21 and graduated in May from William Jewell College in Kansas City. Elizabeth, their daughter, is 19 and is a sophomore at Missouri State University.

Jamie can be reached at fugitt@cofo.edu

MATHEMATICS IN RURAL APPALACHIA

Worksheet 2A

Rodeo Box Plots for Barrel Racing   Name _____

1. Groups which are given one of the tables of **barrel racing** time should complete this worksheet (page 7).

a. Construct a box plot for this data.

b. What is the median time for your data? _____ Where do you find this on the box plot?

c. What is the range of your data? _____ How can you get the range from the box plot?

d. What are the times for the slowest 25% of the riders? _____ Explain how this information is determined from the box plot.

e. What is the interquartile range for this data? _____ How is this obtained from the box plot?

f. Is there a larger range in the fastest 25% of the times or the slowest 25% of the times? _____ How can you determine this from the box plot? Describe why this might be the case.

Wait until the teacher has demonstrated how to use the computer software before working this problem.

Use the software to construct and print a box plot for your data. Compare the computer generated graph to the one you constructed by hand. Explain any differences between the two graphs.

MATHEMATICS IN RURAL APPALACHIA

Worksheet 2B

Rodeo Box Plots for Bull Riding   Name _____

1. Groups which are given one of the tables of **bull riding** scores should complete this worksheet (page 7).

a. Construct a box plot for this data.

b. What is the median score for your data? _____ Where do you find this on the box plot?

c. What is the range of your data? _____ How can you get the range from the box plot?

d. What are the scores for the best 25% of the scores? _____ Explain how this information is determined from the box plot.

e. What is the inter-quartile range for this data? _____ How is this obtained from the box plot?

f. Is there a larger range in the highest 25% of the scores or the lowest 25% of the scores? _____ How can you determine this from the box plot? Describe why this might be the case.

Wait until the teacher has demonstrated how to use the computer software before working this problem.

2. Use the software to construct and print a box plot for your data. Compare the computer generated graph to the one you constructed by hand. Explain any differences between the two graphs.

MATHEMATICS IN RURAL APPALACHIA

Group Project   *Display Your Own Data*   Name _____

This project is to be completed by your group. One set of papers should be submitted for your group. Be sure to include the names of all group members.

Determine some aspect of the Forsyth rodeo which you would like to research for which the data could be represented using a box plot. For example you could research bronco riding scores, calf-roping scores, number of entries in various events over the past several years, amount of income generated by the rodeo over the past several years, number of tickets sold for the past several years, etc. Brainstorm in your group and be creative. If necessary suggest two ideas and let group members vote on which idea to research.

On a 4" x 6" note card explain the idea your group has decided to research. Explain in detail so that someone reading your card can understand exactly what you plan to research. Submit this to the teacher.

Once your research idea has been approved the exciting part begins! You are to pretend that the rodeo officials have hired you to conduct research for them. Your group will be required to gather the data and then do a presentation for the rodeo board. As a minimum, your presentation must include the following printed material.
  Table of the data
  Box plot
  At least 5 general statements about the data. For example, "The middle 50% of the barrel riding times in the Forsyth 2002 rodeo are from about 18.3 seconds to 21.8 seconds."
  All of the information should be typed
  .

Your group presentation should be between 3 and 5 minutes. As a part of your presentation you should explain how your data was obtained and the conclusions which you were able to make. Be sure to indicate how this data is represented on your graph.

References

Department of Elementary and Secondary Education Website. Retrieved on December 20, 2002 at

http://dese.mo.gov/divimprove/curriculum/GLE/MathFinalGLE_3.2.04.pdf.

National Council of Teachers of Mathematics Electronic Illuminations. Retrieved on

December 20, 2002 at http://illuminations.nctm.org/tools/tool_detail.aspx?id=77

National Council of Teachers of Mathematics (2000). *Principles and standards for school mathematics.* Reston, Virginia: National Council of Teachers of Mathematics.

Lesson 18

Farmer Jonah's New Barn

By: Michael Ratliff

Introduction

Nationally, there is a shortage of middle grades mathematics teachers. According to the National Center for Education Statistics, 53.1% of those whose main teaching assignment in middle grades is mathematics do not have the equivalent of either a major or minor in mathematics or mathematics education (CUPM, 2004). This is disheartening given that the middle grades mathematics curriculum is mathematically rich and abundant in rewarding opportunities for teachers.

Nevertheless, the reality is that our middle grades graduates with a mathematics emphasis are teaching in high schools. Therefore, it is important to enhance their experience with mathematics content to better prepare them for a high school teaching assignment.

Considering the aforementioned issues, the rationale for this lesson is twofold: 1) strengthen and reaffirm student skills in the use of geometry software—*Geometer's Sketchpad*® in this case; and 2) better prepare a student who may be teaching in a high school setting by exposing students to more geometry content (i.e., specifically content that could be included in either a middle grades or high school geometry curriculum).

National Content Standards	National Process Standards
Number and Operations: Compute fluently and make reasonable estimates. Algebra: Use mathematical models to represent and understand quantitative relationships. Geometry: Use visualization, spatial reasoning, and geometric modeling to solve problems. Measurement: Understand measurable attributes of objects and the units, systems, and processes of measurement; and Apply appropriate techniques, tools, and formulas to determine measurements. Data Analysis and Probability: N/A www.nctm.org	Problem-Solving: Solve problems that arise in mathematics and other contexts. Reasoning & Proof: Make and investigate mathematical conjectures. Communication: Communicate their mathematical thinking coherently and clearly to peers, teachers, and others Connections: Recognize and apply mathematics in contexts outside of mathematics Representations: Use representations to model and interpret physical, social, and mathematical phenomena www.nctm.org
Kentucky State Standards **Grades 5-9** Students will use properties of circles, arcs, chords, central angles, inscribed angles, and concentric circles. Students will use angle and side relationships such as triangle sum theorem, triangle inequalities, isosceles and equilateral triangle properties, altitude, and median. Students will use angle and side relationships such as triangle sum theorem, triangle inequalities, isosceles and equilateral triangle properties, altitude, and median. Students will use Pythagorean theorem and its converse. Students will use right triangle relationships such as trigonometric ratios (45-45-90 & 30-60-90). http://www.education.ky.gov/	**Your State/District Standards** **(fill in here)**

MATHEMATICS IN RURAL APPALACHIA

Instructional Plan

Objectives	Prior Knowledge
Students successfully completing the lesson will be able to: Develop a solid knowledge (mastery) of geometry at the level of the program certification (5-9) and a level above the highest grade certified (a high school geometry course).	The students are able to know: Compass/straightedge construction of four triangle centers (centroid, circumcenter, incenter, and orthocenter); analogous construction of these triangle centers using *Geometer's Sketchpad®*; and knowledge of measurement tools available in *Geometer's Sketchpad®*.
Discussion Items	**Materials**
Use *Geometer's Sketchpad®* to find where Farmer Jonah should build his new dairy barn? Is volume or surface area a better representation? Why?	Geometer's Sketchpad Map Computer with internet access

The Portion of Farmer Jonah's Farm Located North of County Road #421

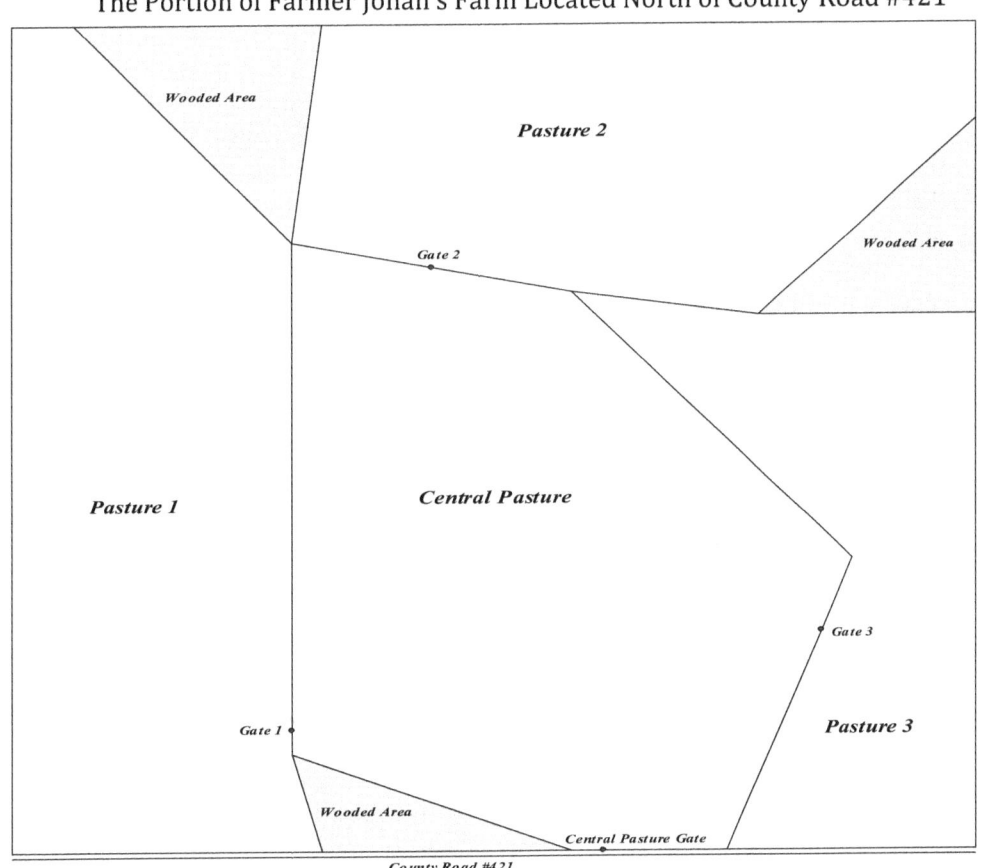

Lesson Procedures

Instructional Sequence & Comments

1. Distribute the problem with diagram and briefly discuss it. (10 minutes)

Comments: Be sure and comment on the use of points for the gates and for the new dairy barn. Also, state that the assumption is that Farmer Jonah's dairy cattle walk in a straight line and that there are no obstacles in the central pasture. Another point to consider is the number of cattle grazing in a pasture. There are two options: (1) State that there is one herd of cattle and that they graze in the same pasture. The grazing pastures are rotated with no pasture used more than any other pasture. (2) State that the same number of cattle graze in each pasture each day. (Note: This issue will be addressed as an extension on the take-home assessment.

2. Instruct students to open the *Sketchpad* file and decide on a method for solving the problem. (5 minutes)

Comments: Students may work independently or in pairs. Regardless of how they choose to work, encourage discussion among them regarding their approaches.

3. Monitor student progress and guide them when necessary. (15 minutes)

Comments: Usually, a student will construct a triangle using the "gate" points as vertices. Then, a random point P is constructed (intuitively, in the interior of the triangle – at some point in the lesson inquire/comment on the point being located in the interior of the triangle) and distances are measured from point P to each of the "gate" points. These distances are then summed. With trial and error, the point P can be repositioned in the diagram such that the sum of the distances to the "gate" points is a minimum – hence, Farmer Jonah's new dairy barn location. (Note: Once one student (or a pair) discovers this method, it spreads to others because of their discussions. Also, encourage students to construct the line segments from point P to the "gate" points. The constructed line segments provide some insight into an important property of point P.)

4. After students have found the new dairy barn location, allow time for students to check point P with triangle centers previously discussed – centroid, circumcenter, incenter, and orthocenter. (10 minutes)

Comments: Students should be proficient in these constructions from a previous lesson. (Give students the option to print their document and construct the centers using a compass and straightedge. Most will choose to use the technology for the constructions.)

5. After students have discovered that point P is none of the other four centers (at least in the given problem – there are cases where the point P will coincide with other triangle centers), allow time for students to investigate properties of point P and develop a method for constructing point P. (15 minutes)

Comments: Closely monitor student progress and guide when necessary.

6. Using a class discussion forum, summarize the methods that students used for constructing point P. (Have students verbally share/demonstrate their methods ("free-handing" on the marker board or using *Sketchpad* with a projector).) (15 minutes)

Comments: When appropriate, comment on student methods. In summarizing, identify point P by its proper names, Fermat's point (also the isogonic point and sometimes referred to as the equiangular point) and comment on Fermat's and Torricelli's methods for constructing the point. Student methods will usually be one or both of these. Refer students to the following website for more information on these methods: http://mathsforeurope.digibel.be/fermat.htm – the significant text from the website follows.

7. Distribute the take-home assessment and briefly discuss it. Emphasize the role of technology in approaching the problem briefly discussing The Technology Principle (PSSM, 2000). (5 minutes)

Comments: The assessment will require students to transfer three points (representing communities) from a map to *Sketchpad*, find the equiangular point, transfer the equiangular point back to the map, and then identify the location. The transferring of the information from the map to *Sketchpad* is content from a previous lesson. Also, extensions to the problem are given for students to consider – a connection to the concept of equilibrium (physics) can be made with the second extension.

Assessment:

Use *The Geometer's Sketchpad®* software to solve the following problem.

The Rural Water Association of South-central Kentucky has decided to build a new water tower to provide water for three communities: Clementsville, Knifley, and Pellyton. (See the map below of the Green River Lake area of Kentucky.) Suppose you are the surveyor contracted to decide where the tower should be located in order that the sum of its distances from the three cities is the smallest, thus providing the most efficient system. Present and justify your decision.

Your Ideas/Comments/Notes for Next Year

About the Author

Michael Ratliff teaches mathematics at Lindsey Wilson College (Columbia, Kentucky). He is pursuing a doctorate in mathematics education through the University of Tennessee and. Ratliff is interested in the mathematical content knowledge needed for teaching geometry at the middle and high school levels. He is also interested in lesson study groups and place-based mathematics education in rural settings.

Lesson 19

How Much Will My Chicken Farm Cost?

By: Sharilyn Owens

Photo courtesy of the U.S. Department of Agriculture

Introduction

Chicken farmers have three different options in raising chickens for market. They can produce pullets, breeders, or broilers. Pullets are only harvested about twice a year, breeders are harvested after 65 days and broilers can be harvested as many as five times per year. However, the Breeder houses are more labor intensive, which must be considered in calculating costs

Some producers also offer bonuses for exceptional production. Our producer provides the feed and immunizations. The farmer is expected to keep these expenses to a minimum. If the farmer uses more feed than is expected, s/he is docked in pay, but if less is used than expected, a bonus is paid.

Chickens that are cold require more feed. If the farmer is willing to run the heat in the houses so the chickens do not get stressed, they will save on feed, and receive more bonuses but energy expenses will be higher. A farmer raising breeders may also receive guaranteed pay based on the average of the last three flocks.

Another interesting factor is the livability rate which for an entire crop is 96%. The mortality rate is about 1% the first week of production and decreases over a 56 day period (for a broiler crop) to a total of about 4%.

NCTM Content Standards	NCTM Process Standards
Number and Operations: Compute fluently and make reasonable estimates. Algebra: Use mathematical models to represent and understand quantitative relationships. Geometry: Use visualization, spatial reasoning, and geometric modeling to solve problems. Measurement: Understand measurable attributes of objects and the units, systems, and processes of measurement; and Apply appropriate techniques, tools, and formulas to determine measurements. Data Analysis and Probability: Data Analysis and Probability: Data Analysis and Probability: Find, use, and interpret measures of center and spread, including mean and interquartile range; discuss and understand the correspondence between data sets and their graphical representations, especially histograms, stem-and-leaf plots, box plots, and scatter plots www.nctm.org	Problem-Solving: Solve problems that arise in mathematics and other contexts. Reasoning & Proof: Make and investigate mathematical conjectures. Communication: Communicate their mathematical thinking coherently and clearly to peers, teachers, and others Connections: Recognize and apply mathematics in contexts outside of mathematics Representations: N/A www.nctm.org
West Virginia State Standards 11th –12th grade Solve application problems using linear, quadratic and exponential functions with emphasis on data collection and analysis. Use appropriate formulas to solve workplace problems calculate costs, simple and compound interest, finance charges, loan payments and taxes. Compare various methods of investing money. http://wvde.state.wv.us/csos/	**Your State/District Standards (fill in here)**

MATHEMATICS IN RURAL APPALACHIA

Instructional Plan

Objectives	Prior Knowledge
Students successfully completing the lesson will be able to: - Estimate most effective cost of building chicken houses ; - Estimate the yearly income based on the number (and type- later discussion) of chick houses built.	The students are able to: - Measure with standard and metric units. - Perform basic operations with whole numbers, fractions, and decimals - Find the layout of the maximum number - Interpret the data
Discussion Items	**Materials**
- How much will it cost me to build a chicken farm? - How do I go about getting a loan? How much can I expect to make? How do I figure out my guaranteed income? - What is the estimated income for the houses I have built on my land? - 4. How many 500, 450 or 400 foot houses can I put on my property?	- Data provided by a local chicken producer - Measurement instrument

MATHEMATICS IN RURAL APPALACHIA

Lesson Procedures

Have students measure their land (or grandparents' land, or neighbors land—if none is available, give them an arbitrary plot of land) and make a scale drawing of it. Then determine the layout of the maximum number of chicken houses, allowing 48 feet between houses, which they can build on the land, and figure out the cost. Some of the costs, such as the road, the gravel, the dead bird disposal, and the generator, need not be determined for every house if one road can service more than one house, and a well can service all of the houses. But, every house needs a wiring and plumbing, so these types of things should be considered when determining cost.

Your Ideas/Comments/Notes for Next Year

	PER (500 foot) HOUSE
PAD & ROAD & GRAVEL	$ 15,700
WELL & PLUMBING TO HOUSE	$ 5,000
BUILDING	$ 77,000
WIRING & PLUMBING	$ 9,000
EQUIPMENT (varies)	$ 61,000
GENERATOR	$ 6,500
DEAD BIRD DISPOSAL	$ 2,500
TOTAL PROJECT COST (for grower)	**$176,700**

	42 X 400	42 X 450	42 X 500
Chicks placed	18,300	20,500	22,800
livability	96%	96%	96%
Birds moved	17,568	19,680	21,888
condemn			
Birds sold	17,568	19,680	21,888
Avg Wt	7.10	7.10	7.10
Pounds sold	124,733	139,728	155,405
Flocks / YR	5.00	5.00	5.00
Pounds sold / YR	623,664	698,640	777,024
Base Pay	$0.0480	$0.0480	$0.0480
Bonus Pay	$0.0060	$0.0060	$0.0060
Utility Pay	$0.0020	$0.0020	$0.0020
Ttl avg pay/lb.	$0.0560	$0.0560	$0.0560
Gross Pay	$34,925.18	$39,123,84	$43,513.34

About the Author

Sharilyn presently teaches math at Wilkes Community College in Wilkesboro NC, a small town in the foothills of the Blue Ridge Mountains, and is pursuing a doctorate degree in the ACCLAIM program. Her rural teaching experience includes 8 years at East Wilkes High School, 2 years at Macon County High School, Montezuma, GA, and 1 year at Fort Valley Middle School, Fort Valley GA. She received her Bachelor's and Master's degrees from University of Georgia.

She also loves cooking and raising her brilliant children and swims for exercise. She listens to all kinds of music, but her all-time favorite is the Eagles. She has a quirky habit of looking for number patterns on the road in her odometer and other people's license plates. She enjoys hunting, fishing and camping, but will never go without running water or toenail polish. She knows Spanish and sign language. She has dual citizenship in Canada and US.

Lesson 20

Confidence Intervals in Achievement Data

By: Paula Schlesinger

Introduction

Indicators of school-level achievement such as the percentage of students who are proficient in a particular content area, are subject to random year-to-year variation in much the same way that the results of an opinion poll will vary from one random sample to another. This random variation, which is more pronounced for a small school, should be taken into account by education officials when evaluating school progress in a policy climate of high stakes. To do otherwise is to unnecessarily risk the false identification of a failing school. In this monograph,

Twenty seven states included confidence intervals in their NCLB accountability plans. The ambitious agenda of the No Child Left Behind Act of 2001 (NCLB 2002) sets unprecedented challenges for public schools in the United States. And these challenges are particularly daunting for states that have a sizable rural population, where the major precepts of NCLB are often at variance with the reality of rural education (e.g., Reeves, 2003; Tompkins, 2003).

To be sure, NCLB provisions regarding school choice, teacher qualifications, technical assistance, supplemental education services, and the evaluation of adequate yearly progress will be tough for any school to accommodate. But these provisions will be considerably more difficult for schools that are small and geographically isolated—that is, for the many schools in this country that reside in rural communities.

http://www.ruraledu.org/docs/nclb/tcsummary.pdf

NCTM Content Standards	NCTM Process Standards
Number and Operations: Compute fluently and make reasonable estimates. Algebra: Use mathematical models to represent and understand quantitative relationships. Geometry: N/A Measurement: N/A Data Analysis and Probability: Find, use, and interpret measures of center and spread, including mean and interquartile range; discuss and understand the correspondence between data sets and their graphical representations, especially histograms, stem-and-leaf plots, box plots, and scatter plots www.nctm.org	Problem-Solving: Solve problems that arise in mathematics and other contexts. Reasoning & Proof: Make and investigate mathematical conjectures. Communication: Communicate their mathematical thinking coherently and clearly to peers, teachers, and others Connections: Recognize and apply mathematics in contexts outside of mathematics Representations: Use representations to model and interpret physical, social, and mathematical phenomena www.nctm.org
North Carolina State Standards Understand & construct confidence intervals for: - Single proportions (using Z). - Single means (using Z). - Single means (using t distribution - Mean differences from paired samples (using t). - Differences between two proportions (using Z). - Differences between two independent means (using Z). - Differences between two independent means (using t). - The slope of the least squares line (using t). http://www.ncpublicschools.org/curriculum/mathematics/scos/1998/09apstat	**Your State/District Standards (fill in here)**

Instructional Plan

Objectives	Prior Knowledge
Students successfully completing the lesson will be able to: - Calculate the confidence interval for a proportion - Explain the means of having a 95% level of confidence - Understand the concept of normal distribution - Touch the normal approximation of the binomial	The students are able to: - Know the concept of proportion - Collect data, interpret data, and analyze data; know the concept of normal distribution.
Discussion Items	**Materials**
- What it means to have a 95% level of confidence? - What does this have to do with a normal distribution? - How many students would a school need to have in grade 8 in order to have a margin of error of less than or equal to +/-1% at a 95% level of confidence?	- Computer - Graphing calculator

MATHEMATICS IN RURAL APPALACHIA

Lesson Procedures

1. Read the article that is found at this URL: http://www.ruraledu.org/docs/nclb/tcsummary.pdf

2. Go to the North Carolina Department of Public Instruction Website http://abcs.ncpublicschools.org/abcs/ and look up the ABC Reports for a k-8 school and a middle school in your county. What proportion of 8th graders at each school was at or above grade level in mathematics? What ABC designation did the two schools achieve for 2002-2003?

3. When we calculate the confidence interval for a proportion, we are asked to first verify if the required assumptions are satisfied. Why is this important

4. Are the required assumptions satisfied by the schools you are studying? The formula for calculating a confidence interval for a proportion with a sample that does not meet the criteria for a normal approximation of the binomial is outside the scope of this course. Here is a website that does these calculations for you: http://home.ubalt.edu/ntsbarsh/Business-stat/otherapplets/ConfIntPro.htm

5. Calculate a 95% confidence interval for the proportion of students scoring at or above grade level for grades 8 for each school. Show your calculations. If you used the website above for any of your calculations explain why.

6. Explain what it means to have a 95% level of confidence. What does this have to do with a normal distribution?

7. What is the margin of error for each school? Why is there a difference in the margin of error between the two schools?

8. How many students would a school need to have in grade 8 in order to have a margin of error of less than or equal to +/-1% at a 95% level of confidence?

9. If you used the upper limit of the confidence intervals to determine the ABC designation for the two schools, what designation would they have achieved?

10. What is the difference between using a confidence interval and using the point estimate provided by the test scores in any given year for determining the ABC designation for schools?

11. What did you learn about confidence intervals by doing this project?

Assessment

Ask each student to present their results in class to indicate how they arrived at the answer: which commands you used in your calculator-or include a print-out of what you did on the computer.

Your Ideas/Comments/Notes for Next Year

About the Author

Paula Schlesinger is a doctoral student in the ACCLAIM program. She teaches mathematics and German at Mayland Community College in Spruce Pine, NC. Paula grew up on a farm in northern Indiana, and has four children and one grandchild.

Paula can be reached at psch53@hotmail.com

Chapter Four: Organizing and Reorganizing

We would like readers to look at these lessons as examples of what they might create by considering how to utilize what they *already know* about everyday rural life to enhance and improve mathematics teaching and learning. This section of the book should help you confront some of the difficulties, especially those related to getting started.

The great cognitive psychologist Jerome Bruner wrote that any subject could be taught to people of any age in an intellectually responsible manner. That is exactly the spirit of adventure that math teachers need in order to connect their classroom to the local people and places. When setting out on this sort of adventure, it is good to get the lay of the land and make some plans. Accordingly, the remainder of this chapter offers some advice towards that end, including some additional sources and plans.

Organization

Much research shows definitively how mathematics education is conducted in America's schools, including America's rural schools. Math in America is most commonly taught with a focus on (a) definitions and (b) routine procedures. Up to the secondary level, math is remains primarily focused on arithmetic. Moreover, the typical curriculum maintains a strong emphasis on memorizing basic computational facts, practice using the facts, and performing a variety of more or less routine calculations in well-structured artificial contexts. At the secondary level, students are often sorted into either a no-new-math track or and "advanced" math track. Many students in the "advanced" math track, however, do not learn enough in high school math to avoid being required to take remedial courses.

These are simply the facts. What they mean is something else again. In the context of this book, we are proposing an infusion of "place" into the mathematics curriculum. This infusion is valuable everywhere but we believe critical for rural America.

Mathematics is so lovely and so useful; it seems genuinely puzzling that so many students should leave our classrooms hating it.

The most prominent organization of mathematics teachers and teacher educators (NCTM, the National Council of Teachers of Mathematics) advises a very, very different approach to mathematics education from what typically happens in math classrooms. The NCTM promotes teaching that focuses on concepts and problem solving and, at the same time, quite literally endorses equity as the first principle. The values of this reform movement are taught in colleges of education and in prominent professional development programs across America.

The movement is meeting with limited success and (as one might expect) daunting resistance. Even the mild notion of an "integrated" mathematics sequence is regarded by the public and by NCTM opponents as zany. *Algebra I, Geometry,* and *Algebra II* must be preserved! Few people realize that other nations have long practiced integrated "maths" (plural). In Britain, one studies "maths" and integration is the norm through grade 10.

Reorganizing

No one truly "knows" what place-based rural mathematics education "is." Place-based rural mathematics is a work in progress, a process that is a constant invention. We learn what it is by making it. This is very different from the ready-made descriptions and standards from state departments of education, textbooks, and national professional organizations.

Developing mathematics lessons with connections to everyday rural life, should enhance student learning and address the standards imposed on you by your district, state, or nation. Our authors have studied the research, considered their own practice and created lessons which attempt to connect students, mathematics and community.

We recommend the following works as helpful to classroom teachers at all levels who wish to put the power of "place" to work in their classrooms.

- Hass, T. & Nachtigal, P. (1998). *Place Value: An Educator's Guide to Good Literature on Rural Lifeways, Environments, and Purposes of Education.*

    Hass and Nachtigal introduce 42 excellent works about rural life and its connections to educational purpose. Also available online at ACCLAIM's Research Clearinghouse: http://www.acclaim-math.org/PVhome.aspx

- Hutchinson, D. (2004). *A Natural History of Place in Education.* NY: Teachers College Press.

    Hutchinson, an environmental educator, examines the idea of place in several ways and describes various ways the idea is being used in schools today.

- Kleinfeld, J., McDiarmid, G. W., & Parrett, W. (1992) *Inventive Teaching: The Heart of the Small School.* Fairbanks, AK: Center for Cross-Cultural Studies.

    Chapter 2 of this book gives clear and practical suggestions for connecting to local communities (using local talent, heritage projects, student service to communities, and student enterprises). Available free, online from ERIC. You'll need the ERIC—ED349153—for your search at http://www.eric.ed.gov/

- Lipka, J. with Mohatt, G., &Ciulistet Cultural Group. (1998). *Transforming the Culture of Schools: Yup'ik Eskimo Examples.* Mahwah, NJ: Lawrence Erlbaum Associates

    Lipka and colleagues have a particular interest in mathematics instruction. Their approach is based more on culture than place, but the idea of community connection is the point.

- Sobel, D. (2004). *Place-based Education: Connecting Classrooms and Communities*. Great Barrington, MA: The Orion Society.

    Sobel provides a host of practical examples; including many dealing with mathematics instruction (Sobel covers the main school subjects). ACCLAIM staff reviewed both this book and the Hutchinson volume: http://www.acclaim-math.org/docs/html_rme/rem8/03.03fea_nichols_howley_theorizing.html .

There are additional web-based resources including full-text access to the archives of *Rural Mathematics Educator*, our online journal:

http://www.acclaim-math.com/newsletters.aspx

For additional ideas and news about connecting to communities from the organization most active in promoting rural schools and communities, go to the Rural School and Community Trust's website: http://www.ruraledu.org/. You can subscribe to their electronic newsletter for teachers and community people at the Rural Roots archive:

http://www.ruraledu.org/site/apps/nl/content.asp?c=beJMIZOCIrH&b=1164615&ct=2891745.

www.ingramcontent.com/pod-product-compliance
Lightning Source LLC
Chambersburg PA
CBHW081233170426
43198CB00017B/2749